NEW DEVELOPMENTS IN MEDICAL RESEARCH

HUMAN PAPILLOMAVIRUSES (HPVS)

DETECTION, INFECTION AND PROTECTION

NEW DEVELOPMENTS IN MEDICAL RESEARCH

Additional books and e-books in this series can be found on Nova's website under the Series tab.

NEW DEVELOPMENTS IN MEDICAL RESEARCH

HUMAN PAPILLOMAVIRUSES (HPVS)

DETECTION, INFECTION AND PROTECTION

YVETTE FRANKLIN
EDITOR

nova
Medicine & Health
New York

Library of Congress Cataloging-in-Publication Data

ISBN: 978-1-53617-331-4

Published by Nova Science Publishers, Inc. † New York

CONTENTS

PREFACE

Virtually all cervical cancer cases are caused by human papilloma viruses (HPV), mainly HPV-16 and HPV-18. Furthermore, HPVs may cause other types of cancer, such as squamous anal cancers and vulvar cancer. In men, HPVs are associated with genital lesions, anal cancer, penile cancer, and head and neck cancers. As such, this compilation focuses on recent findings in HPV epidemiology, prevention, and updates regarding prophylatic HPV vaccines.

Bioinformatics tool are important in the analysis of molecular and biological data, which could provide novel insights on the development of treatment and prevention methods for disease control. Through bioinformatics, it is possible to analyze and simulate HPV molecular data through the use of universal databases and analysis tools, in order to better understand the virus genetics and infection cycle. In conclusion, Human Papillomaviruses (HPVs): Detection, Infection and Protection appraises HPV vaccines for the prophylaxis of HPV infection in healthy individuals. Vaccines used for earlier varieties are Gardasil (Merck, Kenilworth, NJ, U.S.A.) and Cervarix (GlaxoSmithKline, Brentford, U.K.).

Chapter 1 - Cervical cancer (CC) is the third most frequently diagnosed cancer in women in the world with an estimated 569,847 new cases and 311,365 deaths per year, which places it as the third cause of cancer death in women worldwide with Human Papilloma Viruses (HPVs) being the main etiological agent of CC and premalignant lesions. Virtually all cervical cancer cases are caused by HR-HPV, mainly HPV-16 and HPV-18 which are associated to 70% of all cervical cancer cases. Furthermore, HPVs may cause other types of cancer, such as squamous anal cancers, and vulvar cancer. In men, HPVs are associated with genital lesions, anal cancer, penile cancer, and head and neck cancers.

HPVs are circular, double-stranded DNA viruses of ~8 kb in size, that belong to the *Papillomaviridae* family, and can be classified according to their epithelial tropism in cutaneotropic or mucosotropic. HPVs genome contain approximately 8 ORFs that are divided into three functional regions: *early region* (E) that codes for the viral proteins (E1-E7) which are essential for viral replication; a *late region* (L) that codes for the capsid proteins L1 and L2; and a *long region of control* (LCR) of ~ 1 kb that contains regulatory elements involved in the replication and transcription of viral DNA. The infected basal cells are human keratinocytes and form the reservoir of infection. In these cells, the viral genome is maintained in a low copy number as an episome. As these cells divide, they produce daughter cells that are pushed outwards towards the epithelial surface. Different events in the virus life cycle take place at different stages during this migration.

The main risk for the development of anogenital cancer is a persistent infection with HR-HPV. Therefore, primary and secondary prevention measures and practice become crucial to avoid HPV-associated cancer. HPV vaccines can prevent infection against some types of HPVs that cause cancer and genital warts. Furthermore, the main goal of screening is to identify precancerous lesions caused by HPV in order to remove it, and to prevent the development of invasive cancers. Until recently, cytology-based screening, known as the Pap test, was the only method for the detection of HPV infection and HPV-assocted lessions. Nowadays, additional tests are available with the advent of higly specific and sensitive molecular tests for HPV.

With respect to HPV-associated cancers and genital warts prevention, the FDA approved three commercially available HPV vaccines that have been proven to be both effective and safe: Gardasil (against types 6, 11, 16, 18), Cervarix (against types 16 and 18) and Gardasil 9 (9-valent Vaccine).

The three HPV vaccine targets the HPV types that cause cervical but also vulva, vagina, anus, and oropharynx cancer. Gardasil 4 and 9 also protects against the HPV types that cause most genital warts (types 11 and 6). This review will focus on recent findings on HPV epidemiology, prevention of infection and updates regarding prophylatic HPV vaccines and the newer therapeutic approach still on development. New HPV screening approaches are also reviewed.

Chapter 2 - Human papillomavirus (HPV) is the most common sexually transmitted infection capable of causing papillomas and cancer. For example, HPV is the main cause of cervical cancer. Currently, prevention strategies are associated with the use of prophylactic vaccines, and the diagnosis is performed by cytopathological exams, such as Papanicolau. The treatment of HPV-related lesions is associated with the removal of infected areas. However, the available vaccines only protect against a small number of HPV types, meaning that some high-risk HPV types might also infect vaccinated people. In addition, as far as we know, there is still no available effective drugs against the virus. Therefore, novel innovative strategies are needed in order to help the development of new treatment and prevention methods for HPV control. In this context, Bioinformatics tools arise as an important approach to analyze molecular and biological data, which could bring novel insights and facilitates the development of treatment and prevention methods for disease control. Through Bioinformatics, it is possible to analyze and simulate HPV molecular data through the use of universal databases and analysis tools, in order to better understand the virus genetics and infection cycle, as well as to identify possible targets for the development of therapeutical drugs and vaccines. There are several Bioinformatics softwares that could be used for therapeutical target identification for HPV control. Within silico analysis, it is possible to identify and calculate the expression of human genes that might be involved in a metabolic pathway associated with HPV infection, providing insights for drug predictions. Rational drug design of specific inhibitors against the virus and host proteins is a promising strategy for the development of novel therapies. By combining Bioinformatics tools with in vitro assays, it is possible to drastically reduce time and costs associated with development of drugs against cervical cancer and other HPV-related diseases. In addition, Bioinformatics could also be used to help the identification of conservative immunogenic peptides, which might be able to protect against a broader spectrum of HPV types. In conclusion, despite the available vaccines, HPV-related diseases is still a major worldwide problem, mainly because its association with cervical cancer development. Therefore, the development of novel treatment and prevention methods for HPV control is still of importance, and Bioinformatics emerges as a relevant auxiliary tool in the fight against HPV infection and cervical cancer.

Chapter 3 - HPV [Human Papilloma Virus] is one known cause for inducing oncogenesis. This chapter appraises HPV vaccines for prophylaxis of HPV infection in healthy individuals. Effective vaccines exist to ensure prophylaxis against HIV morbidity and mortality. Inoculation against HPV, to acquire immunity, is advised for all individuals before sexual debut.

In: Human Papillomaviruses (HPVs) ISBN: 978-1-53617-331-4
Editor: Yvette Franklin

Chapter 1

AN UPDATE IN THE DETECTION, PREVENTION, AND TRANSFORMATION POTENTIAL OF HUMAN PAPILLOMAVIRUSES (HPVS)

Jimena Hochmann*, PhD and Felipe Parietti†
Virology Section, Science Faculty, Universidad de la Republica,
Montevideo, Uruguay

ABSTRACT

Cervical cancer (CC) is the third most frequently diagnosed cancer in women in the world with an estimated 569,847 new cases and 311,365 deaths per year, which places it as the third cause of cancer death in women worldwide with Human Papilloma Viruses (HPVs) being the main etiological agent of CC and premalignant lesions. Virtually all cervical cancer cases are caused by HR-HPV, mainly HPV-16 and HPV-18 which are associated to 70% of all cervical cancer cases. Furthermore, HPVs may cause other types of cancer, such as squamous anal cancers, and vulvar

---

* Corresponding Author's Email: jhochmann@fcien.edu.uy.
† Corresponding Author's Email: fparietti@fcien.edu.uy.

cancer. In men, HPVs are associated with genital lesions, anal cancer, penile cancer, and head and neck cancers.

HPVs are circular, double-stranded DNA viruses of ~8 kb in size, that belong to the *Papillomaviridae* family, and can be classified according to their epithelial tropism in cutaneotropic or mucosotropic. HPVs genome contain approximately 8 ORFs that are divided into three functional regions: *early region* (E) that codes for the viral proteins (E1-E7) which are essential for viral replication; a *late region* (L) that codes for the capsid proteins L1 and L2; and a *long region of control* (LCR) of ~ 1 kb that contains regulatory elements involved in the replication and transcription of viral DNA. The infected basal cells are human keratinocytes and form the reservoir of infection. In these cells, the viral genome is maintained in a low copy number as an episome. As these cells divide, they produce daughter cells that are pushed outwards towards the epithelial surface. Different events in the virus life cycle take place at different stages during this migration.

The main risk for the development of anogenital cancer is a persistent infection with HR-HPV. Therefore, primary and secondary prevention measures and practice become crucial to avoid HPV-associated cancer. HPV vaccines can prevent infection against some types of HPVs that cause cancer and genital warts. Furthermore, the main goal of screening is to identify precancerous lesions caused by HPV in order to remove it, and to prevent the development of invasive cancers. Until recently, cytology-based screening, known as the Pap test, was the only method for the detection of HPV infection and HPV-assocted lessions. Nowadays, additional tests are available with the advent of higly specific and sensitive molecular tests for HPV.

With respect to HPV-associated cancers and genital warts prevention, the FDA approved three commercially available HPV vaccines that have been proven to be both effective and safe: Gardasil (against types 6, 11, 16, 18), Cervarix (against types 16 and 18) and Gardasil 9 (9-valent Vaccine). The three HPV vaccine targets the HPV types that cause cervical but also vulva, vagina, anus, and oropharynx cancer. Gardasil 4 and 9 also protects against the HPV types that cause most genital warts (types 11 and 6). This review will focus on recent findings on HPV epidemiology, prevention of infection and updates regarding prophylatic HPV vaccines and the newer therapeutic approach still on development. New HPV screening approaches are also reviewed.

Keywords: cervical cancer, HPV screening, vaccination, infection

1. INTRODUCTION

1.1. Human Papillomavirus (HPV) and Cervical Cancer (CC)

HPVs are small, non-enveloped DNA viruses with over 226 types listed to date on the Papillomavirus Episteme (PaVE) database. HPV genotypes 16 and 18 are responsible for approximately 70% of cervical cancers, with the remaining 30% caused by other carcinogenic HPV types. They belong to the *Papillomaviridae* family, and it is believed that they have co-evolved along with the first amniotes around 350 million years ago [1, 2, 3]. This is why their viral cycle is closely related to the differentiation status of keratinocytes in the epithelium. HPVs are classified into five genera based on their tropism represented by the Greek letters: α, β, γ, μ and ν. In turn, HPVs can be classified according to their epithelial tropism in cutaneotropics (predominantly β and γ genera, μ, ν and some α to a lesser extent) or mucosotropic (exclusively some types within the α genus) [4]. Cutaneotropic HPVs cause hyperproliferative lesions on the skin called warts, especially in the area of the hands and feet. It should be noted that not all HPV infections cause visible clinical manifestations, in most cases the infections are asymptomatic and are rapidly resolved by the host's immune system. On the other hand, mucosotropic HPVs are of greater health relevance than the previous ones.

Cervical Cancer (CC) is the third most frequently diagnosed cancer in women in the world with an estimated 569,847 new cases per year, and 311,365 annual deaths by 2018, which places it as the third cause of death from cancer worldwide in women regarding GLOBOCAN 2018 data. The International Agency for Research on Cancer (IARC) classified 12 α-HPVs as type I carcinogens (types 16, 18, 31, 33, 35, 39, 45, 51, 52, 56, 58, and 59). Almost all cervical cancer cases (CC) are linked to a persistent infection by one or more high-risk HPVs, with types 16 and 18 reporting the highest incidence worldwide [5, 6]. However, low-risk HPVs such as HPV-11 and HPV-6 cause genital warts that are usually benign and resolve after 6 months in 90% of the cases. These same viral types have been linked to the development of recurrent respiratory papillomatosis (PRR), which is a rare but serious illness that mainly affects children under the age of five, and it is characterized by the exacerbated growth of papillomas in the upper respiratory tract, especially in the the larynx obstructing breathing and speech [7].

1.2. HPV Genome

The HPV genome consist of double stranded DNA and encodes for approximately 8 open reading frames (ORFs), which are divided into three regions: an early region (E) that codes for the proteins needed to viral replication and cellular transformation (E1, E2, E4, E5, E6 and E7); a late region (L) that encodes for capsid proteins L1 and L2, and a non-coding region (LCR or URR) that contains sites of binding to viral and cellular transcription factors, the origin of genome replication, and controls gene expression, since there are early (P_E) and late (P_L) promoters [8] (Figure 1).

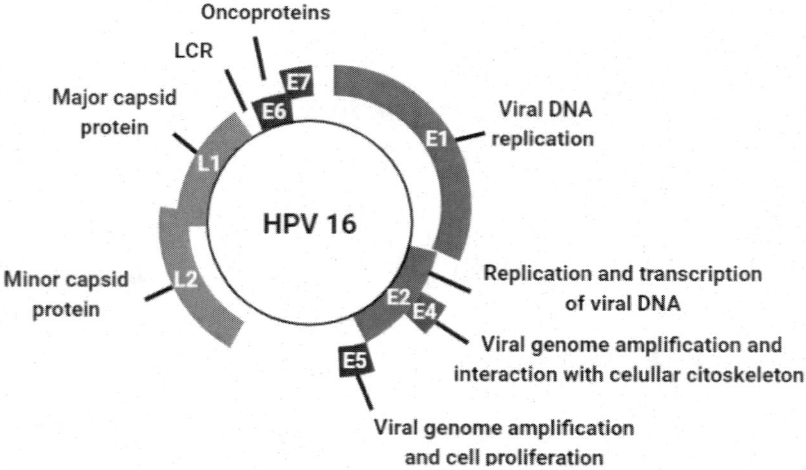

Figure 1. Organization of the Genome of HPV-16 indicating their main viral genes.

L1 encodes for the major capsid protein, while L2 codes for the minor capsid protein that binds to viral DNA and facilitates its packaging within the virion [9]. E1 codes for a viral helicase that has ATPase activity [10]. E1 is able to unwind viral DNA in the origin of replication and recruit the cellular replication machinery, the union of E1 to the origin of replication is stabilized by E2 [10]. E2 is in fact a group of small proteins that regulate the transcription of viral DNA, the replication and partitioning of the E2 genome functions as a repressor of E6/E7 [11]. E4, plays an important role in the escape of the virion from the cornified layers of the epithelium, it is generally expressed as the fusion product E1^E4 and it is the most abundant expressed protein in the upper layers of the epithelium during a productive infection [12].

1.2.1. E6 Oncoprotein

E6 is one of the viral oncoproteins of approximately 16 kDa found mainly in the nucleus, although it also can be found in the cytosol, whose main target is the p53 tumor suppressor protein, which is a DNA site-specific transcription factor, and one of the key signaling coordinators in the cell cycle, following genotoxic and cytotoxic stress, that forms a trimeric complex together with a cellular adapter protein called E6-AP, which is an E3 ubiquitin ligase (E6/E6-AP/p53). E6 protein, binds to the LxxLL consensus sequence in the conserved domain of E6-AP, and as a heterodimer is able to degrade p53 [13]. This complex is ubiquitinized for degradation in the proteasome [14]. The degradation of p53 interferes with the appropriate signaling pathways to repair the damage, or trigger cell death in cases where the damage is beyond repair. This, in turn allows for replication of damaged DNA, and survival of cells with deleterious mutations that would normally be eliminated. E6 proteins of high and low risk HPVs interact with the C-terminus of p53, but high-risk HPVs appear to bind it with a higher affinity, which may explain the oncogenicity of these types [15]. Apart from p53 protein, constitutive expression of E6 of HR-HPVs is able to induce the expression of the catalytic subunit of telomerase (hTERT) (Figure 2). A recent study reported that alpha-species 9 (HPV16 and HPV16-related types), HPV18, HPV39 and HPV51 upregulate the hTERT promoter [16].

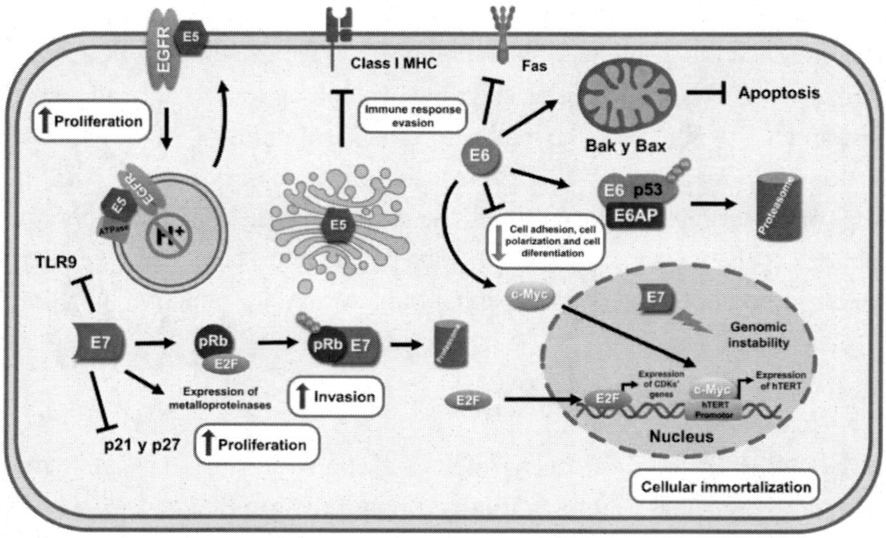

Figure 2. Interaction of the viral oncoproteins E5, E6 and E7 with their main cellular targets.

It is suggested that the activation of hTERT occurs by two different mechanisms: activation of the hTERT promoter by the overexpression of the c-Myc proto-oncogene and the transcription factor Sp1; and on the other hand, by the degradation of hTERT repressor (NFX-91). It has been reported that the telomerase activation and p53 degradation is critical for immortalization of epithelial cells by HPV16 [17, 18]. In addition, E6 also interacts with paxilin, which is a protein involved in focal adhesion, and in actin organization. This interaction subsequently inactivates paxilin causing a disruption in the maintenance of the cell cytoskeleton and in the organization and differentiation of the epithelium [19], with the transcriptional co-activator p300/CBP (CREB-binding protein) [20], and also with other proteins involved in the regulation of transcription and replication of DNA, such as IRF-3 and ADA3. Furthermore, E6 of high and low risk HPVs also binds and speeds up the degradation of pro-apoptosis proteins involved in apoptosis pathways (both extrinsic and intrinsic), such as BAK, the TNF-1 receptor (TNF-R1), and FADD associated with death domain FAS, since E6 binds to the tumor necrosis factor receptor 1 (TNF-R1) preventing the activation of caspases 3 and 8 [21]. Finally it is worthy to mention that another important target of E6 of high-risk HPVs are those proteins that have a PDZ domain that regulate extremely important functions such as polarity and cell motility because the high risk E6 have a PDZ-binding domain at the C-terminus that is absent in E6 of low-risk HPVs [22].

1.2.2. E7 Oncoprotein

HPV16 E7 is a small acidic polypeptide of approximately 13kDa. This protein is located in the cytoplasm [23, 24, 25] but it is also found in nuclear pools. Its main target are the retinoblastoma proteins (pRbs) which phosphorylate E2F proteins, controlling the expression of genes involved in the differentiation, development, cell proliferation, and apoptosis, and this causes the release of the transcription factor E2F (Figure 2). The released E2F factor translocates to the nucleus and recruits different remodeling factors of the chromatin, including deacetylases of histones (HDAC) that promote the expression of genes activated by E2F such as genes of cyclin A and E which leads to an increase in cell proliferation. E7 also inactive the "pocket proteins" p107 and p130 regulators of E2F via the LXCXE motif in CR2 [26, 27, 28]. These pockets proteins have the ability to regulate G1/S entry, and progression by modulating the transcriptional activities of E2F transcription factors.

In addition, high-risk HPV E7 proteins also contribute to cell cycle deregulation through several additional mechanisms such as the inactivation of regulatory proteins of the G1-S transition of the cell cycle, inhibitors of Cdks p21 and p27, this causes the cell proliferation to be deregulated due to the activity of Cdk2.

Recent studies also showed that E7 induces breaks in DNA, while increasing levels of expression of one sirtuin, and an NAD-dependent de-acetylase deacetylase (Sirt1), reduces the levels of histone γ-H2AX, that participates in the repair of DNA double strand breaks [29]. E7 is insuficient to increase endogenous hTERT mRNA or telomerase activity, although it augmented E6-induced hTERT transcription and telomerase activity [30].

1.2.3. E5 Oncoprotein

The E5 oncoprotein has been the least studied at the level of cellular transformation, however, it is known that same as E6 and E7 contributes to the carcinogenic process. E5 is a small transmembrane protein of 8.3 kDa (HPV-18) and hydrophobic that is located mainly in the intracellular membranes of the reticulum endoplasmic (RE) and the Golgi apparatus [31, 32]. It is also able to decrease the acidification of endosomes (from a pH of 5.9 to 6.9) blocking the activity of the vacuolar ATPase, this leads to a receptor retention in the plasma membrane preventing its degradation, the recycling of them in the cell membrane, and the extension of its active signal [33, 34]. In this way E5 is able to stimulate cell proliferation forming complexes with epidermal growth factor (EGFR) in the cell membrane [34, 35] (Figure 2). It has also reported that E5, through its interaction with the Golgi apparatus, is able to reduce expression of MHC class I in the cell membrane; which plays an important role in the evasion of the response immune mediated by cytotoxic T lymphocytes [36].

1.3. Oncogenic Potential of HR- HPV E5, E6, E7

The viral oncoproteins E5, E6 and E7 of high-risk oncogenic HPVs are the main viral factors responsible for malignant transformation and progression to cervical cancer. E6 and E7 are expressed in a constitutive manner after the integration of the viral genome into the genome of the host cell; they are pleiotropic oncoproteins, since they have many targets at the cellular level. The integration of the viral genome, or subgenomic fragments is purely accidental and results in the deregulation of the expression of E6 and E7, since the integration interrupts the open reading frame of E2, the main repressor of the expression of E6 and E7[37]. The constitutive expression of these oncogenes influences directly and indirectly in certain cellular signaling pathways that regulate processes highly regulated such as, cell proliferation, migration and invasion, deregulation of these processes can lead to to the development of carcinogenesis. High levels of expression of E6 and E7 lead to the accumulation of damage in cellular DNA resulting in genomic instability and a reduced responsiveness of DNA repair mechanisms [37]. Cells that express HR E6 and E7 as a product of viral genome integration have a greater proliferative capacity than the cells that express the viral episome. The fact that the loss of the function of E2 is critical for malignant progression was demonstrated by experiments that show that restatement of E2 in cervical cancer cell lines causes suppression of cell growth. These experiments showed that constitutive expression of E6 / E7 is necessary to maintain the phenotype transformed [38].

As for E5 oncoprotein transformation potential, it has a weak transforming activity in the absence of E6 and E7. However, studies conducted on different cell models have demonstrated that HPV-16 E5 is able to cooperate with E6 and E7 significantly in the transformation process. Boulenouar et al. have showed using a "trophoblast type" cells as a model (BeWo) the transformative effects of the three viral oncoproteins together (E5, E6 and E7) of HPV-16 on a placental cell line, reporting an increase in cell proliferation, migration and invasion in HPV-16 E5/E6/E7 cells [39]. These cells are a good study model of the development of carcinogenesis since they have certain characteristics in common with tumor cells such as high proliferative capacity, independent anchor growth and absence of contact inhibition. The results obtained by Boulenouar et al. provided solid evidence that E5 cooperates with the E6 and E7 oncoproteins during the progression to malignancy by increasing invasive capacity, proliferative and migratory of BeWo cells [39]. This work is supported by Al-Moustafa, who suggest that E5 of HPV-16 may cooperates together with E6 and E7 to induce epitelial mesenchymal transition (EMT) via the signaling pathway of epidermal growth factor receptors (EGFR) [40], since another study reported that E6/E7 of HPV type 16 induce EMT and consequently cell motility and invasion via epidermal growth factor receptor (EGFR) signaling pathways [41], and HPV 16 E5 been shown to up-regulates the expression of vascular endothelial growth factor (VEGF) through the activation of EGFR, MEK/ERK1 and 2 as well as PI3K/Akt,11 which could induce cell invasion via EMT initiation [42]. On the other hand, the work of Kivi et al. reported the effect of HPV 16 E5 in the transcriptome of human epithelial cells using cDNA microarrays. In this study, the E5 protein affects several cellular pathways involved in cell adhesion, cell motility and mitogenic signaling [43].

It has also been documented that E5 of HPV-6 and HPV-16 are able to cooperate with E7 in the transformation and immortalization of murine primary cells [44].

1.4. HPV Infection

Infections by the oncogenic HPV genotypes 16, 18, 31, 33 and the low-risk HPV11 are the best studied HPVs at the cellular and molecular levels. Host cell entry of HPV is initiated by binding of the virus particle to cell surface receptors, and then, these particles interact with their natural host cell, basal keratinocytes, throught via the glycosaminoglycan (GAG), specially Heparan sulfate proteoglycan (HSPGs) [45, 46]. It is well known that HPVs bind to a widely expressed and evolutionary conserved cell surface receptor, and the interaction depends mainly by L1 [47]. HSPGs are frequently found in the extracellular matrix (ECM), and on the cell surface of adherent cells [48]. As it was mentioned before, HPV particles entry appears to occur specifically in the basal keratinocytes of the mucosal epithelium subsequent to the binding of virions to the basement membrane epithelium. It has been also observed that a secondary cell adhesion receptor (a6-integrin), is involved in cell to cell interactions, and also in the infectious internalization of HPV subsequent to interaction with HSPG [49]. It semms that HSPG functions as more than a simple attachment factor in the HPV infection, since this interaction also promotes essential conformational changes in the viral capsid.

So, it is believed that a productive infection would require an initial low specificity binding mediated by L1 protein, followed by the interaction of a more specific protein component with L2. In the capsid, most of L2 is protected, although there is a portion of the amino terminal L2 region exposed on the surface. During infection, the virus initially binds to heparin sulfate proteoglycans in the extracellular matrix through L1. Next, L1 undergoes a conformational change that leaves L2 exposed. Later on the cell surface, cellular cyclophilins bind to L2 and expose to amino-terminal portion containing a furin cleavage site [50]. L2 is cleaved by furin on the cell surface at a consensus cleavage site that is conserved among all papillomaviruses. After L2 cleavage, the virion enters the cell through a secondary receptor that mediates internalization by endocytosis mediated by clathrin and / or caveolin-coated vesicles, depending on the type infectious viral particle. The virion then travels through the endosomal system to the Golgi apparatus. During this process, acidification of L1 and L2 proteins occurs, provoking the dissociation of the two proteins. Finally, the complex of L2 along the viral genome is transferred to the nucleus where it remains in episomal form during the productive life cycle [50].

1.5. Detection Methods to Identify HPVs Presence in Clinical Samples

Persistent infection with HR-HPV is the main risk factor necessary but not sufficient for cervical cancer, which develops over a long period of time through precursor lesions, which can be detected by cytological screening. The majority of these lesions regress spontaneously without treatment (90%). The importance of screening is to detect the lesions as soon as posible, before they progress to a malignant lesion [51]. The most employed technique to detect cervical premalignant lesion is the cytology Papanicolau Test, although it has decreased the incidence of cervical cancer, HPV-related cervical disease, continues to represent a major problem on health-care systems. These observations could be explained in part due to low specificity of screening tests, which could give a significant variability in the diagnosis of cervical dysplastic lesions. For these reason, nowadays molecular tests are used in conjuction with the Pap-test. These molecular tests consist in detecting the HPV genome in clinical specimens, based on nucleic probe technology [52]. HPV diagnosis relies on molecular-biology techniques that allow its accurate detection and typing [53]. At present two methods of diagnosis are available, signal amplification assays and target-acid amplification.

Regarding signal-amplification assays, they rely on *in situ* hybridization techniques in liquid phase which consist on denaturing target DNA (HPV), and incubating the sample with labeled probes which bind mainly to the L1 region of the HPV genome. These assays consist of comercial kits such as DigeneW HPV test using Hybrid CaptureW 2 (hc2) technology, and the CervistaW HPV HR assay. These are the oldest tests validated by the Food Drug Administration (FDA) in the United States which can distinguish between HR and LR groups, but they do not identify different genotypes of HPV [54, 55]. In the case of The CervistaW HPV (Hologic, Inc., Marlborough, MA, USA) kit, it detects the presence of 14 HR-HPV types (16,-18,-31,-33,-35,-39,-45,-51,-52,-56,-58,-59,-66 and -68) [56].

The target acid-amplification methods consist of techniques of gene amplification by Polymerase Chain Reaction (PCR), which require a low amount of target DNA initially. This is a sensitive and relatively fast method, in which PCR amplifies only the genomic sequence of a single HPV type. The PCR employs several pairs of specific primers in the same reaction (multiplex PCR) to amplify fragments of different sizes by genotype. These methods include some conventional and commercial techniques: Papillochek, PCR, RT-PCR, PCR-RFLP, Linear Array, HPV Genome Sequencing, INNOLIPA, and Luminex-based HPV genotyping. These methods are extremely sensitive, and allow multiplex analysis. As for PapilloCheck® (Greiner Bio One), it can detect and genotype 24 HPV types in a single reaction (HPV 6, 11, 16, 18, 31, 33, 35, 39, 40, 42, 43, 44, 45, 51, 52, 53, 55, 56, 58, 59, 66, 68, 70, 73, and 82). This assay uses a multiplex PCR with fluorescent primers to amplify a 350 bp fragment of the E1 gene of HPV. Then it carries out a hybridization of amplicons obtained with specific probes immobilized on chips: each HPV type is detected by a specific probe, which is present on each chip in five copies. The bound DNA is then marked with a fluorescent molecule which is excited at a given wavelength [57, 58]. This method allows the identification and amplification of multiple infections in one clinical simple. Other techniques such as Linear Array Genotyping Test, consist on co-amplifying of a 450 bp region of L1 and the β-globine gene by PCR with biotinylated consensus primers PGMY09/11 and PC04/GH20 respectively, coupled with a reverse line-blot hybridization, which then will be denatured and hybridized with specific probes immobilized on strips. This assay allows the discrimination of 37 HPV, including 15 HR types, 16, 18, 31, 33, 35, 39, 45, 51, 52, -56, 58, 59, 68,73 and -82), 3 probable HR (26, 53 and 66), and 10 LR (6, 11, 40, 42, 54, 61, 70,72, 81 and -CP6108) and 9 genotypes for which the risk is still undetermined (55, 62, 64, 67, 69, 71, 83, 84 and -IS39) [59].

A study compared the clinical performance of HC2 with the PapilloCheck® for the detection of underlying CIN2 or CIN3 using cervical samples from a study with a randomized population-based trial of 24,510 women undergoing routine cervical screening in the United Kingdom National Health Service Cervical Screening program (NHSCSP). This study suggested that HR PapilloCheck test could be a suitable HPV detection assay for use in the cervical screening [60]. Furthermore, another study compared the effectiveness of the two human papillomavirus (HPV) genotyping techniques, Linear Array and PapilloCheck, in women with high-grade squamous intraepithelial lesion (HSIL) observing that both methods showed equivalent performance for the detection of oncogenic HPV types in women with HSIL, however PapilloCheck test seems to have the advantage of being a method that avoids bias when reading the HPV genotypes [61].

The INNOLiPA, test is based on the same principle: the extracted DNA is amplified by PCR with biotinylated consensus primers SPF10 that amplify 65 bp region of the HPV L1 gene, and the 270 bp of the human HLA-DP1 gene using SPF10 biotinylated primers followed by genotyping The genotypes identified are determined by comparison with the check strip. This test detects 28 HPV genotypes: 18 high-risk HPV, seven low-risk HPV, and three unclassified [60, 62]. Actually this assay has undergone several modifications, resulting in a few different versions. The INNO-LiPA HPV Genotyping Extra II assay (INNO-LiPA; Fujirebio Europe, Ghent, Belgium), target 32 types, four types more than the previous versión. This actual version contains genotype specific probes for more decisive genotyping results, an upgraded SPF10 primer set resulting in improved sensitivity (comparable for all hrHPV types), improved human DNA control primers and provides a ready-to-use amplification reagent.

As for the HPV-PCR methods, these use consensus, and degenerate primers such as PGMY09/PGMY1 and GP5+/GP6+, which allow amplification of a large number of HPV genotypes in a single reaction. The primers PGMY09/11of 20 bp target the conserved region of the L1 capsid gene leading to an amplicon of 450 bp. After amplification, the HPV genotypes can be determined separately, using diverse techniques such as genotyping by restriction-fragment length polymorphism (RFLP), which allows the HPV to be typed, and is less expensive than sequencing. Furthermore, HPV genotypes can be determined by direct sequencing using the conventional Sanger technique.

The Luminex-based HPV genotyping, is a sensitive, reproducible technique for the simultaneous genotyping of 27 High risk genotypes of HPV, and is a method based on polymerase chain reaction (PCR) using primers GP5+/50 biotinylated GP6+ to detect the different genotypes [63]. This technology consists in using polystyrene, or magnetic microbeads coupled with specific oligonucleotide probes, that are internally dyed with red and infrared fluorophores of differing intensities. In multiplex immunoassay kits, individual bead sets are then coated with a capture biotinilated antibody. Multiple analyte-specific beads can then be combined in a single well of a 96-well microplate-format assay to detect and quantify multiple targets simultaneously, using one of the Luminex® instruments for analysis. In this case, after amplification of HPV DNA with fluorescent primers, the amplicons are denatured then hybridized on dyed microbeads labeled with fluorophores of each genotype specific HPV [64, 65]. Different studies have showed that Cervista had a high concordance with Luminex technology (96.26%), since both methods shared a similar clinical value in the detection of CIN2 or higher [66].

1.6. HPV Vaccination

Since its introduction in June 2006, HPV vaccination has been steadily gaining ground [67]. As of 2019, 115 countries and territories around the world have introduced an HPV vaccination program as part of its immunization schedule with 20 of them and 4 territories implementing a gender-neutral approach and over 270 million vaccine doses distributed during the last decade.. An additional 39 countries plan to implement HPV vaccination by 2021 [68]. Currently there are three different HPV vaccines available in the market based on virus-like particles (VLPs) from the major capsid protein L1. The quadrivalent vaccine Gardasil, which confers immunity against types 6, 11, 16 and 18 respectively, the bivalent vaccine Cervarix that protects against types 16 and 18. And more recently, the nonavalent vaccine Gardasil 9 targeting types 6, 11, 16, 18, 31, 33, 45, 52, and 58. All of the current vaccines available are prophylactic and not therapeutic which means they do not target infected cells. The three vaccines provide similar immunogenicity and capability for the prevention of cervical cancer, primarily caused by HPV16 and HPV18. Moreover, several observational studies have suggested a high effectiveness of HPV vaccination in reducing the prevalence of vaccine-related and non-vaccine related genotypes and preventing cervical intraepithelial neoplasia of grade 2 and 3 (CIN 2 or CIN 3) [68, 69].

1.6.1. Therapeutic HPV Vaccine: E6 and E7 Epitopes as Possible Target Candidates

Unlike prophylactic vaccines aimed at the formation of virus neutralizing antibodies, therapeutic vaccines elicit a T-cell cytotoxicity response leading to the clearance of malignant cells expressing viral oncoproteins. The ideal targets for vaccine immunotherapy are highly conserved HR HPV oncoproteins E6 and E7 expressed in precancerous and tumor tissues [70, 71]. Therapeutic vaccination could be one of the potentially effective treatments for HPV-associated cancer. Some studies have showed that during a persistent infection, the immune response is aimed at E6 and E7 epitopes, making them ideal candidates for the development of a therapeutic vaccine. A challenge that must be overcome in the development of an efficient therapeutic vaccines is that CD4+ T cells which recognize E6 of various HPV types are rare, with only highly conserved parts of E6 being cross-recognized [72].

1.6.2. Vaccine Safety

The three HPV vaccines are safe to administer with only minor side effects such as pain and swelling in the injection site being the most common cause of medical consultation. Some allergic reactions have been observed, along with syncope in vaccinated girls. Extensive media coverage was given to events of girls recently vaccinated against HPV reporting headache, paresthesia of the lower or upper limbs, respiratory distress, chest pain and fainting [67, 73]. The media reports linked the events to HPV vaccination, which led to strong negative public reaction, reduced acceptance of the HPV vaccine and in the case of Peru, temporarily halted the program's implementation. However, there is not strong evidence suggesting its link with more serious adverse effects such as Guillain-Barré syndrome according to the latest report of the Global Advisory Committee for the Safety of Vaccines (GACVS), the committee has evaluated that the risk of developing Guillain-Barré as a result of vaccination is not bigger than 1 case per 1.000.000 people vaccinated. Another study also points towards the safety of aluminum adjuvants pointing out that it has been estimated that the rate of ASIA syndrome is 3.6 cases per 100,000 doses of anti-HPV vaccine, thus deeming the vaccine safe for use [74].

CONCLUSION

In conclusion, this chapter summarizes our current knowledge on HPV virology. Although many advances have been made, there are still fundamental questions to be addressed regarding the malignant transformation of HPV and the development of therapeutical vaccines that could enhance the immune response to fight off cancer cells.

REFERENCES

[1] Gottschling, M., A. Stamatakis, I. Nindl, E. Stockfleth, A. Alonso, and I. G. Bravo. (2007). "Multiple Evolutionary Mechanisms Drive

Papillomavirus Diversification." *Molecular Biology and Evolution,* 24 (5): 1242–58. https://doi.org/10.1093/molbev/msm039.

[2] Shah, S.., Doorbar, J., and Goldstein. R. (2010). "Analysis of Host–Parasite Incongruence in Papillomavirus Evolution Using Importance Sampling." *Molecular Biology and Evolution,* 27 (6): 1301–14. https://doi.org/10.1093/molbev/msq015.

[3] Van Doorslaer, Koenraad. (2013). "Evolution of the Papillomaviridae." *Virology,* 445 (1–2): 11–20. https://doi.org/10.1016/j.virol.2013.05.012.

[4] de Villiers, E. M., Fauquet, C., Broker, T., Bernard, H. U., and zur Hausen., H. (2004). "Classification of Papillomaviruses." *Virology,* 324 (1): 17–27. https://doi.org/10.1016/j.virol.2004.03.033.

[5] Walboomers, J. M. M., Jacobs, M., Manos, M., Xavier, F., Bosch, J. Kummer, A., Shah, K., Snijders, P., Peto, J., Meijer, C., and Muñoz., N. (1999). "Human Papillomavirus Is a Necessary Cause of Invasive Cervical Cancer Worldwide." *The Journal of Pathology,* 189 (1): 12–19. https://doi.org/10.1002/(sici)1096-9896(199909)189:1<12::aid-path431>3.0.co;2-f.

[6] Berois, N., Nozar, F., Greif, D., Franciulli, A., Barrios, E., and Osinaga, E. (2017). "Prevalence and Distribution of High-Risk Human Papillomavirus Genotypes in Invasive Carcinoma of the Uterine Cervix in Uruguay. An Update on Clinical Outcome." *Medical Research Archives,* 5 (6). https://doi.org/10.18103/mra.v5i6.1303.

[7] Major, T. (2005). "The Characteristics of Human Papillomavirus DNA in Head and Neck Cancers and Papillomas." *Journal of Clinical Pathology,* 58 (1): 51–55. https://doi.org/10.1136/jcp.2004.016634.

[8] Morshed, K., Polz-Gruszka, D., Szymański, M., and Polz-Dacewicz., M. (2014). "Human Papillomavirus (HPV) – Structure, Epidemiology and Pathogenesis." *Otolaryngologia Polska,* 68 (5): 213–19. https://doi.org/10.1016/j.otpol.2014.06.001.

[9] Wang, J. W., and Richard B.S. Roden. (2013). "L2, the Minor Capsid Protein of Papillomavirus." *Virology,* 445 (1–2): 175–86. https://doi.org/10.1016/j.virol.2013.04.017.

[10] Bergvall, M., Melendy, T., and Archambault, J. (2013). "The E1 Proteins." *Virology,* 445 (1–2): 35–56. https://doi.org/10.1016/j.virol.2013.07.020.

[11] McBride, Alison A. (2013). "The Papillomavirus E2 Proteins." *Virology,* 445 (1–2): 57–79. https://doi.org/10.1016/j.virol.2013.06.006.

[12] Doorbar, J. (2013). "The E4 Protein; Structure, Function and Patterns of Expression." *Virology,* 445 (1–2): 80–98. https://doi.org/10.1016/j.virol.2013.07.008.

[13] Martinez-Zapien, D., Ruiz, X. F., Poirson, J., Mitschler, A., Ramirez, J., Forster, A., Cousido-Siah, A. et al. (2016). "Structure of the E6/E6AP/P53 Complex Required for HPV-Mediated Degradation of P53." *Nature,* 529 (7587): 541–45. https://doi.org/10.1038/nature16481.

[14] Scheffner, Martin, Jon M. Huibregtse, Richard D. Vierstra, and Peter M. Howley. (1993). "The HPV-16 E6 and E6-AP Complex Functions as a Ubiquitin-Protein Ligase in the Ubiquitination of P53." *Cell*, 75 (3): 495–505. https://doi.org/10.1016/0092-8674(93)90384-3.

[15] Lechner, M. S., D. H. Mack, A. B. Finicle, T. Crook, K. H. Vousden, and L. A. Laimins. (1992). "Human Papillomavirus E6 Proteins Bind P53 in Vivo and Abrogate P53-Mediated Repression of Transcription." *The EMBO Journal,* 11 (8): 3045–52. https://doi.org/10.1002/j.1460-2075.1992.tb05375.x.

[16] Van Doorslaer, Koenraad, and Robert D. Burk. (2012). "Association between HTERT Activation by HPV E6 Proteins and Oncogenic Risk." *Virology,* 433 (1): 216–19. https://doi.org/10.1016/j.virol.2012.08.006.

[17] Gewin, L., and D. A. Galloway. (2001). "E Box-Dependent Activation of Telomerase by Human Papillomavirus Type 16 E6 Does Not Require Induction of c-Myc." *Journal of Virology,* 75 (15): 7198–7201. https://doi.org/10.1128/jvi.75.15.7198-7201.2001.

[18] Klingelhutz, Aloysius J., Foster, S., and McDougall, J. (1996). "Telomerase Activation by the E6 Gene Product of Human Papillomavirus Type 16." *Nature,* 380 (6569): 79–82. https://doi.org/10.1038/380079a0.

[19] Tong, X., and P. M. Howley. (1997). "The Bovine Papillomavirus E6 Oncoprotein Interacts with Paxillin and Disrupts the Actin Cytoskeleton." *Proceedings of the National Academy of Sciences,* 94 (9): 4412–17. https://doi.org/10.1073/pnas.94.9.4412.

[20] Patel, D., S. M. Huang, L. A. Baglia, and D. J. McCance. (1999). "The E6 Protein of Human Papillomavirus Type 16 Binds to and Inhibits Co-Activation by CBP and P300." *The EMBO Journal,* 18 (18): 5061–72. https://doi.org/10.1093/emboj/18.18.5061.

[21] Filippova, M, T A Brown-Bryan, C A Casiano, and P J Duerksen-Hughes. (2005). "The Human Papillomavirus 16 E6 Protein Can Either Protect or Further Sensitize Cells to TNF: Effect of Dose." *Cell Death & Differentiation,* 12 (12): 1622–35. https://doi.org/10.1038/sj.cdd.4401678.

[22] Songyang, Z., A. S. Fanning, C. Fu, J. Xu, S. M. Marfatia, A. H. Chishti, A. Crompton, A. C. Chan, J. M. Anderson, and L. C. Cantley. (1997). "Recognition of Unique Carboxyl-Terminal Motifs by Distinct PDZ Domains." *Science,* 275 (5296): 73–77. https://doi.org/10.1126/science.275.5296.73.

[23] Huh, K.-W., J. DeMasi, H. Ogawa, Y. Nakatani, P. M. Howley, and K. Munger. (2005). "Association of the Human Papillomavirus Type 16 E7 Oncoprotein with the 600-KDa Retinoblastoma Protein-Associated Factor, P600." *Proceedings of the National Academy of Sciences,* 102 (32): 11492–97. https://doi.org/10.1073/pnas.0505337102.

[24] Nguyen, Christine L., McLaughlin-Drubin, M., and Münger, K. (2008). "Delocalization of the Microtubule Motor Dynein from Mitotic Spindles by the Human Papillomavirus E7 Oncoprotein Is Not Sufficient for Induction of Multipolar Mitoses." *Cancer Research,* 68 (21): 8715–22. https://doi.org/10.1158/0008-5472.can-08-1303.

[25] Smotkin, D., and F. O. Wettstein. (1987). "The Major Human Papillomavirus Protein in Cervical Cancers Is a Cytoplasmic Phosphoprotein." *Journal of Virology,* 61 (5): 1686–89. https://jvi.asm.org/content/61/5/1686.

[26] Greenfield, I., J. Nickerson, S. Penman, and M. Stanley. (1991). "Human Papillomavirus 16 E7 Protein Is Associated with the Nuclear

Matrix." *Proceedings of the National Academy of Sciences,* 88 (24): 11217–21. https://doi.org/10.1073/pnas.88.24.11217.

[27] Dyson, N. P., K. Guida, K. Münger, and E. Harlow. (1992). "Homologous Sequences in Adenovirus E1A and Human Papillomavirus E7 Proteins Mediate Interaction with the Same Set of Cellular Proteins." *Journal of Virology,* 66 (12): 6893–6902. https://jvi.asm.org/content/66/12/6893.long.

[28] Münger, K., B.A. Werness, N. Dyson, W.C. Phelps, E. Harlow, and P.M. Howley. (1989). "Complex Formation of Human Papillomavirus E7 Proteins with the Retinoblastoma Tumor Suppressor Gene Product." *The EMBO Journal,* 8 (13): 4099–4105. https://doi.org/10.1002/j.1460-2075.1989.tb08594.x.

[29] Allison, S. J., Ming J., and Milner, J. (2009). "Oncogenic Viral Protein HPV E7 Up-Regulates the SIRT1 Longevity Protein in Human Cervical Cancer Cells." *Aging,* 1 (3): 316–27. https://doi.org/10.18632/aging.100028.

[30] Liu, X., Roberts, J., Dakic, A., Zhang, Y., and Schlegel., R. (2008). "HPV E7 Contributes to the Telomerase Activity of Immortalized and Tumorigenic Cells and Augments E6-Induced HTERT Promoter Function." *Virology,* 375 (2): 611–23. https://doi.org/10.1016/j.virol.2008.02.025.

[31] Conrad, M., V. J. Bubb, and R. Schlegel. (1993). "The Human Papillomavirus Type 6 and 16 E5 Proteins Are Membrane Associated Proteins Which Associate with the 16-Kilodalton Pore-Forming Protein." *Journal of Virology,* 67 (10): 6170–78. https://jvi.asm.org/content/67/10/6170.long.

[32] Schlegel, R, M Wade-Glass, M. Rabson, and Y. Yang. (1986). "The E5 Transforming Gene of Bovine Papillomavirus Encodes a Small, Hydrophobic Polypeptide." *Science,* 233 (4762): 464–67. https://doi.org/10.1126/science.3014660.

[33] Disbrow, G. L., J. A. Hanover, and R. Schlegel. (2005). "Endoplasmic Reticulum-Localized Human Papillomavirus Type 16 E5 Protein Alters Endosomal PH but Not Trans-Golgi PH." *Journal of Virology,* 79 (9): 5839–46. https://doi.org/10.1128/jvi.79.9.5839-5846.2005.

[34] Crusius, K., Auvinen, E., Steuer, V., Gaissert, H., and Alonso, A. (1998). "The Human Papillomavirus Type 16 E5-Protein Modulates Ligand-Dependent Activation of the EGF Receptor Family in the Human Epithelial Cell Line HaCaT." *Experimental Cell Research,* 241 (1): 76–83. https://doi.org/10.1006/excr.1998.4024.

[35] Crusius, K., I. Rodriguez, and A. Alonso. (2000). "The Human Papillomavirus Type 16 E5 Protein Modulates ERK1/2 and P38 MAP Kinase Activation by an EGFR-Independent Process in Stressed Human Keratinocytes." *Virus Genes,* 20 (1): 65–69. https://link.springer.com/article/10.1023%2FA%3A1008112207824.

[36] Campo, M. S., S. V. Graham, M. S. Cortese, G. H. Ashrafi, E. H. Araibi, E. S. Dornan, K. Miners, C. Nunes, and S. Man. (2010). "HPV-16 E5 down-Regulates Expression of Surface HLA Class I and Reduces Recognition by CD8 T Cells." *Virology,* 407 (1): 137–42. https://doi.org/10.1016/j.virol.2010.07.044.

[37] Munger, K., A. Baldwin, K. M. Edwards, H. Hayakawa, C. L. Nguyen, M. Owens, M. Grace, and K. Huh. (2004). "Mechanisms of Human Papillomavirus-Induced Oncogenesis." *Journal of Virology,* 78 (21): 11451–60. https://doi.org/10.1128/jvi.78.21.11451-11460.2004.

[38] Münger, K., W. C. Phelps, V. Bubb, P. M. Howley, and R. Schlegel. (1989). "The E6 and E7 Genes of the Human Papillomavirus Type 16 Together Are Necessary and Sufficient for Transformation of Primary Human Keratinocytes." *Journal of Virology,* 63 (10): 4417–4421. https://jvi.asm.org/content/63/10/4417.

[39] Boulenouar, S., C. Weyn, M. Van Noppen, M. Moussa Ali, M. Favre, P. O. Delvenne, F. Bex, A. Noel, Y. Englert, and V. Fontaine. (2009). "Effects of HPV-16 E5, E6 and E7 Proteins on Survival, Adhesion, Migration and Invasion of Trophoblastic Cells." *Carcinogenesis,* 31 (3): 473–80. https://doi.org/10.1093/carcin/bgp281.

[40] Moustafa, Ala-Eddin Al. (2015). "E5 and E6/E7 of High-Risk HPVs Cooperate to Enhance Cancer Progression through EMT Initiation." *Cell Adhesion & Migration,* 9 (5): 392–93. https://doi.org/10.1080/19336918.2015.1042197.

[41] Jung, Y. S., Kato, I. and Kim., H. R. (2013). "A Novel Function of HPV16-E6/E7 in Epithelial–Mesenchymal Transition." *Biochemical and Biophysical Research Communications,* 435 (3): 339–44. https://doi.org/10.1016/j.bbrc.2013.04.060.

[42] Kim, S. -H., Y. -S. Juhnn, S. Kang, S. -W. Park, M. -W. Sung, Y. -J. Bang, and Y. -S. Song. (2006). "Human Papillomavirus 16 E5 Up-Regulates the Expression of Vascular Endothelial Growth Factor through the Activation of Epidermal Growth Factor Receptor, MEK/ ERK1,2 and PI3K/Akt." *Cellular and Molecular Life Sciences,* 63 (7–8): 930–38. https://doi.org/10.1007/s00018-005-5561-x.

[43] Kivi, N, D Greco, P Auvinen, and E Auvinen. (2007). "Genes Involved in Cell Adhesion, Cell Motility and Mitogenic Signaling Are Altered Due to HPV 16 E5 Protein Expression." *Oncogene,* 27 (18): 2532–41. https://doi.org/10.1038/sj.onc.1210916.

[44] Valle, G. F., and L. Banks. (1995). "The Human Papillomavirus (HPV)-6 and HPV-16 E5 Proteins Co-Operate with HPV-16 E7 in the Transformation of Primary Rodent Cells." *Journal of General Virology,* 76 (5): 1239–45. https://doi.org/10.1099/0022-1317-76-5-1239.

[45] Giroglou, T., L. Florin, F. Schafer, R. E. Streeck, and M. Sapp. (2001). "Human Papillomavirus Infection Requires Cell Surface Heparan Sulfate." *Journal of Virology,* 75 (3): 1565–70. https://doi.org/ 10.1128/jvi.75.3.1565-1570.2001.

[46] Joyce, G., Tung, J., Przysiecki, C., Cook, J., Lehman, J., Sands, K., Keller, P. (1999). "The L1 Major Capsid Protein of Human Papillomavirus Type 11 Recombinant Virus-like Particles Interacts with Heparin and Cell-Surface Glycosaminoglycans on Human Keratinocytes." *Journal of Biological Chemistry,* 274 (9): 5810–22. https://doi.org/10.1074/jbc.274.9.5810.

[47] Roden, R. B., R. Kirnbauer, A. B. Jensson, D. R. Lowy, and J. T. Schiller. (1994). "Interaction of Papillomaviruses with the Cell Surface." *Journal of Virology,* 68 (11): 7260–7266. https://www.ncbi. nlm.nih.gov/pmc/articles/PMC237166/.

[48] Spillmann, D. (2001). "Heparan Sulfate: Anchor for Viral Intruders?" *Biochimie,* 83 (8): 811–17. https://doi.org/10.1016/s0300-9084(01)01290-1.

[49] Day, P. M., D. R. Lowy, and J. T. Schiller. (2008). "Heparan Sulfate-Independent Cell Binding and Infection with Furin-Precleaved Papillomavirus Capsids." *Journal of Virology,* 82 (24): 12565–68. https://doi.org/10.1128/jvi.01631-08.

[50] Wiens, Mayim E., and Jason G. Smith. (2014). "Alpha-Defensin HD5 Inhibits Furin Cleavage of Human Papillomavirus 16 L2 To Block Infection." Edited by M. J. Imperiale. *Journal of Virology,* 89 (5): 2866–74. https://doi.org/10.1128/jvi.02901-14.

[51] Termini, L., and LL. Villa. (2008). "Biomarkers in Screening of Cervical Cancer." *Jornal Brasileiro de Doenças Sexualmente Transmissiveis,* 20 (02): 125–31.

[52] Villa, L., and Denny., L. (2006). "Methods for Detection of HPV Infection and Its Clinical Utility." *International Journal of Gynecology & Obstetrics,* 94 (1): S71–80. https://doi.org/10.1016/s0020-7292(07)60013-7.

[53] Shen-Gunther, J., and Yu., X. (2011). "HPV Molecular Assays: Defining Analytical and Clinical Performance Characteristics for Cervical Cytology Specimens." *Gynecologic Oncology,* 123 (2): 263–71. https://doi.org/10.1016/j.ygyno.2011.07.017.

[54] Emmadi, R., J. B. Boonyaratanakornkit, R. Selvarangan, V. Shymala, B. L. Zimmer, L. Williams, B. Bryant, et al. (2011). "Molecular Methods and Platforms for Infectious Diseases Testing a Review of FDA-Approved and Cleared Assays." *The Journal of Molecular Diagnostics,* 13 (6): 583–604. https://doi.org/10.1016/j.jmoldx.2011.05.011.

[55] Arney, A., and Bennett, K. (2010). "Molecular Diagnostics of Human Papillomavirus." *Laboratory Medicine,* 41 (9): 523–30. https://doi.org/10.1309/lm75wgjsvmi7vvef.

[56] Einstein, Mark H., Mark G. Martens, Francisco A. R. Garcia, Daron G. Ferris, Amy L. Mitchell, Stephen P. Day, and Marilyn C. Olson. (2011). "Clinical Validation of the Cervista HPV HR and 16/18

Genotyping Tests for Use in Women With ASC-US Cytology." *Obstetrical & Gynecological Survey,* 66 (4): 211–12. https://doi.org/10.1097/ogx.0b013e318225cd2e.

[57] Dalstein, V., Merlin, S., Bali, C., Saunier, M., Dachez, R. and Ronsin, C. (2009). "Analytical Evaluation of the PapilloCheck Test, a New Commercial DNA Chip for Detection and Genotyping of Human Papillomavirus." *Journal of Virological Methods,* 156 (1–2): 77–83. https://doi.org/10.1016/j.jviromet.2008.11.002.

[58] Crosbie, E., Bailey, A., Sargent, A., Gilham, C., Peto, J., and Kitchener, H. (2015). "The PapilloCheck Assay for Detection of High-Grade Cervical Intraepithelial Neoplasia." Edited by A. M. Caliendo. *Journal of Clinical Microbiology,* 53 (11): 3553–59. https://doi.org/10.1128/jcm.01578-15.

[59] Melchers, W., J. Bakkers, J. Wang, P. de Wilde, H. Boonstra, W. Quint, and A. Hanselaar. (1999). "Short Fragment Polymerase Chain Reaction Reverse Hybridization Line Probe Assay to Detect and Genotype a Broad Spectrum of Human Papillomavirus Types." *The American Journal of Pathology,* 55 (5): 1473–1478. https://doi.org/10.1016/S0002-9440(10)65462-4.

[60] Poljak, M., and Boštjan J. K. (2010). "Commercially Available Assays for Multiplex Detection of Alpha Human Papillomaviruses." *Expert Review of Anti-Infective Therapy,* 8 (10): 1139–62. https://doi.org/10.1586/eri.10.104.

[61] Serravalle, K., J. E. Levi, K. Oliveira, C. Queiroz, A. Dantas, and E. Studar. (2015). "Comparison of Two Techniques for HPV Genotyping in Women with High-Grade Squamous Intraepithelial Lesion." *Revista Brasileira de Ginecologia e Obstetrícia,* 37 (2): 94–99. https://doi.org/10.1590/SO100-720320150005206.

[62] Kleter, B., van Doorn, L. J., Schrauwen, L et al. (1999). "Development and clinical evaluation of a highly sensitive PCR-reverse hybridization line probe assay for detection and identification of anogenital human papillomavirus." *J. Clin. Microbiol*, 37(8): 2508–2517.

[63] Geraets, D. T., K. Cuschieri, M. N. C. de Koning, L. J. van Doorn, P. J. F. Snijders, C. J. L. M. Meijer, W. G. V. Quint, and M. Arbyn.

(2014). "Clinical Evaluation of a GP5+/6+-Based Luminex Assay Having Full High-Risk Human Papillomavirus Genotyping Capability and an Internal Control." *Journal of Clinical Microbiology,* 52 (11): 3996–4002. https://doi.org/10.1128/jcm.01962-14.

[64] Oh, Y., Su, M., Kim, Y. W., Choi, H. S., Nam, G. H., Han, S. J., Hak Park, C., Cho, Y., Han, and Woong, B. D., Ahn, S. (2007). "Polymerase Chain Reaction-Based Fluorescent Luminex Assay to Detect the Presence of Human Papillomavirus Types." *Cancer Science,* 98 (4): 549–54. https://doi.org/10.1111/j.1349-7006.2007.00427.x.

[65] Ozaki, S., Kato, K., Abe, Y., Hara, H., Kubota, H., Kubushiro, K., Kawahara, E., and Inoue, M. (2014). "Analytical Performance of Newly Developed Multiplex Human Papillomavirus Genotyping Assay Using Luminex XMAP^{TM} Technology (Mebgen^{TM} HPV Kit)." *Journal of Virological Methods,* 204 (August): 73–80. https://doi.org/10.1016/j.jviromet.2014.04.010.

[66] García, Dabeiba A. (2011). "Highly Sensitive Detection and Genotyping of HPV by PCR Multiplex and Luminex Technology in a Cohort of Colombian Women with Abnormal Cytology." *The Open Virology Journal,* 5 (1): 70–79. https://doi.org/10.2174/1874357901105010070.

[67] Brotherton, Julia M. L., and Paul N. Bloem. (2018). "Population-Based HPV Vaccination Programmes Are Safe and Effective: 2017 Update and the Impetus for Achieving Better Global Coverage." *Best Practice & Research Clinical Obstetrics & Gynaecology,* 47 (February): 42–58. https://doi.org/10.1016/j.bpobgyn.2017.08.010.

[68] Sanjosé, S de, M Brotons, DS LaMontagne, and L Bruni. (2019). "Human Papillomavirus Vaccine Disease Impact beyond Expectations." *Current Opinion in Virology,* 39 (18): 16–22. https://doi.org/10.1016/j.coviro.2019.06.006.

[69] Kavanagh, Kimberley, Kevin G Pollock, Kate Cuschieri, Tim Palmer, Ross L Cameron, Cameron Watt, Ramya Bhatia, et al. (2017). "Changes in the Prevalence of Human Papillomavirus Following a National Bivalent Human Papillomavirus Vaccination Programme in Scotland: A 7-Year Cross-Sectional Study." *The Lancet Infectious*

Diseases, 17 (12): 1293–1302. https://doi.org/10.1016/s1473-3099(17)30468-1.

[70] Latsuzbaia, Ardashel, Marc Arbyn, Steven Weyers, and Joël Mossong. (2018). "Human Papillomavirus Vaccination Coverage in Luxembourg – Implications of Lowering and Restricting Target Age Groups." *Vaccine,* 36 (18): 2411–16. https://doi.org/10.1016/j.vaccine.2018.03.054.

[71] Ma, Barbara, Bharat Maraj, Nam Phuong Tran, Jayne Knoff, Alexander Chen, Ronald D Alvarez, Chien-Fu Hung, and T.-C. Wu. (2012). "Emerging Human Papillomavirus Vaccines." *Expert Opinion on Emerging Drugs,* 17 (4): 469–92. https://doi.org/10.1517/14728214.2012.744393.

[72] Hende, Muriel van den, Anke Redeker, Kitty M. C. Kwappenberg, Kees L. M. C. Franken, Jan W. Drijfhout, Jaap Oostendorp, A. Rob P. M. Valentijn, et al. (2010). "Evaluation of Immunological Cross-Reactivity between Clade A9 High-Risk Human Papillomavirus Types on the Basis of E6-Specific CD4+Memory T Cell Responses." *The Journal of Infectious Diseases* 202 (8): 1200–1211. https://doi.org/10.1086/656367.

[73] Luciani, S, L Bruni, I Agurto, and C Ruiz-Matus. n. d. "HPV Vaccine Implementation and Monitoring in Latin America." *Salud Publica de Mexico* 60 (6): 683–92. https://doi.org/10.21149/9090.

[74] Cervantes, Jorge L, and Amy Hoanganh Doan. (2018). "Discrepancies in the Evaluation of the Safety of the Human Papillomavirus Vaccine." *Memórias Do Instituto Oswaldo Cruz,* 113 (8). https://doi.org/10.1590/0074-02760180063.

In: Human Papillomaviruses (HPVs)
Editor: Yvette Franklin
ISBN: 978-1-53617-331-4

Chapter 2

BIOINFORMATICS AS AN AUXILIARY TOOL FOR THE TREATMENT AND PREVENTION OF HUMAN PAPILLOMAVIRUS INFECTION

Marcus Vinicius de Aragão Batista*[*], *Gerlane dos Santos Barros*, *Maria da Conceição Viana Invenção*, *Lucas Alexandre Barbosa de Oliveira Santos*, *Adriana de Oliveira Fernandes*, *Everton Silva Mota*, *Claudio Ramos Moreira Filho*, *Jânio Almeida de Souza* and *Luiz Fernando de Jesus Nascimento

Laboratory of Molecular Genetics and Biotechnology,
Department of Biology, Center for Biological and Health Sciences,
Federal University of Sergipe, São Cristóvão, Sergipe, Brazil

* Corresponding Author's Email: mvabatista@hotmail.com.

Abstract

Human papillomavirus (HPV) is the most common sexually transmitted infection capable of causing papillomas and cancer. For example, HPV is the main cause of cervical cancer. Currently, prevention strategies are associated with the use of prophylactic vaccines, and the diagnosis is performed by cytopathological exams, such as Papanicolau. The treatment of HPV-related lesions is associated with the removal of infected areas. However, the available vaccines only protect against a small number of HPV types, meaning that some high-risk HPV types might also infect vaccinated people. In addition, as far as we know, there is still no available effective drugs against the virus. Therefore, novel innovative strategies are needed in order to help the development of new treatment and prevention methods for HPV control. In this context, Bioinformatics tools arise as an important approach to analyze molecular and biological data, which could bring novel insights and facilitates the development of treatment and prevention methods for disease control. Through Bioinformatics, it is possible to analyze and simulate HPV molecular data through the use of universal databases and analysis tools, in order to better understand the virus genetics and infection cycle, as well as to identify possible targets for the development of therapeutical drugs and vaccines. There are several Bioinformatics softwares that could be used for therapeutical target identification for HPV control. Within silico analysis, it is possible to identify and calculate the expression of human genes that might be involved in a metabolic pathway associated with HPV infection, providing insights for drug predictions. Rational drug design of specific inhibitors against the virus and host proteins is a promising strategy for the development of novel therapies. By combining Bioinformatics tools with in vitro assays, it is possible to drastically reduce time and costs associated with development of drugs against cervical cancer and other HPV-related diseases. In addition, Bioinformatics could also be used to help the identification of conservative immunogenic peptides, which might be able to protect against a broader spectrum of HPV types. In conclusion, despite the available vaccines, HPV-related diseases is still a major worldwide problem, mainly because its association with cervical cancer development. Therefore, the development of novel treatment and prevention methods for HPV control is still of importance, and Bioinformatics emerges as a relevant auxiliary tool in the fight against HPV infection and cervical cancer.

Keywords: HPV, bioinformatics, drugs, vaccine

HPV: A Public Health Concern and Its Challenges

Human papillomavirus (HPV) is a circular double-stranded DNA virus that causes a variety of conditions including warts and multiple malignancies involving the squamous epithelium [1]. HPV is the etiological agent of cervical cancer, the fourth most common type of cancer in women worldwide [2]. HPV infections are frequent and constitute a serious public health problem, especially in low- to middle-income countries [3, 4]. It is the most common sexually transmitted disease, with studies estimating an amount of 50% of both men and women worldwide being infected at least once in their lifetime [5]. Due to its transient nature, the immune system is generally able to halt the disease progression and revert the symptoms [6]. However, the main problem lies on the virus type. HPV types are divided in high-risk types, which causes malignant lesions, and low-risk types, which cause benign warts. HPV types 16 and 18 are the most associated to carcinogenesis, and they happen to be the most commonly detected in the whole world [7]. HPV types 6 and 11 are the most prevalent low-risk types in the world. There are over 200 types of HPV described in the Papillomavirus Episteme Database (PaVE) (https://pave.niaid.nih.gov/), and almost 40 types preferentially infect the epithelial cells of the genital tract [1, 8-9]. In this context, the need for cervical cancer prevention strategies is evident.

The main strategy for preventing HPV infection is the use of prophylactic vaccines. Currently, there are three approved vaccines available for use, the 9-valent HPV vaccine (9vHPV), which contains HPV 6, 11, 16, 18, 31, 33, 45, 52 and 58 virus-like particle (VLP); the quadrivalent HPV vaccine (4vHPV), which contains HPV 6, 11, 16, and 18 VLPs, and the bivalent HPV vaccine (2vHPV), which contains HPV 16 and 18 VLPs. These vaccines are generally made by using L1 proteins in the form of VLPs [10]. Although having a good coverage, the efficacy of these vaccines decreases in people who already had contact with the virus, which is very likely due to its global distribution, especially in least developed regions [11]. It is worth mentioning that the 9-valent vaccine is not available in all countries. For that reason, until today, HPV is the seventh common cause of cancer, with 528.000 new cases detected in 2012 and almost 266.000 deaths worldwide [12].

Aside from the previously mentioned reasons, cervical cancer related to HPV is still frequent due to the limitations of the diagnostic methods. The diagnosis is based on the identification of cellular changes along with the identification of viral DNA by molecular biology methods. However, in many countries, especially in South America, Southeast Asia and Africa, patients have to rely on screening methods, such as Papanicolaou (Pap smear) and visual inspection with acetic acid (VIA) [13], which are highly unspecific regarding the virus type, progression and other molecular characteristics. These regions also portray a considerable number of health illiteracy [14], which makes women inconsistent in partaking in these tests at least two times a year. In this context, accurate HPV diagnosis and typing is very important for the infection prevention.

These challenges are what moves scientists to further comprehend and try to develop new strategies towards HPV diagnosis and treatment. An interesting methodology that has arisen in the past decades is bioinformatics, which comprises knowledge from natural and computer sciences. Bioinformatics have proven to be an optimal resource for making treatment/diagnosis design faster, more specific and less expensive. Therefore, in the following sections we will clarify how this technology could be used in order to assist the development of novel HPV diagnostic and treatment methods.

Bioinformatics: A Tool That Can Help Human Papillomavirus Researches

Large-scale data is difficult to calculate when observed in nature [15]. With technology advances, large amounts of data are generated with a single experiment, which highlighted the importance of the use and development of computational methods to analyze these data, becoming the subject called Bioinformatics.

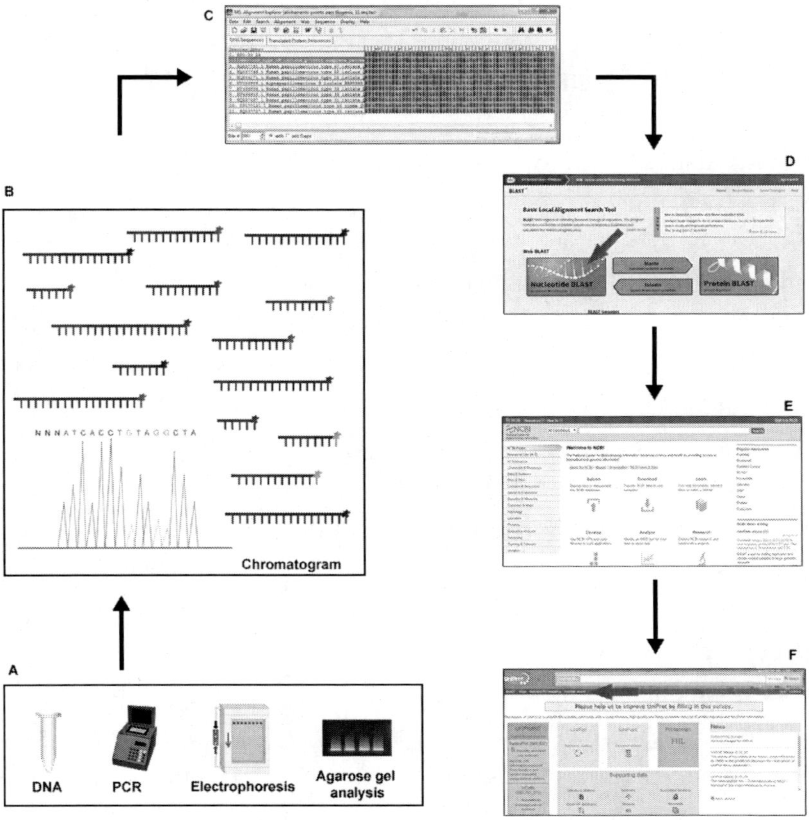

Figure 1. An example of a Bioinformatics workflow since the genetic material extraction (A) until sequence comparison in databases (E and F). The process starts with the in vitro analysis carried out through DNA extraction, amplification (PCR) and visualization of the amplification result by electrophoresis (A). Then, sequencing is performed with the amplified and purified genetic material (B) [21]. From that, multiple sequence alignment is carried out using an alignment algorithm, which are incorporated in MEGA software (C) [17]. BLAST can be used to compare sequences against a specified database, allowing the identification of the sequences in the sample (D) [18], which could be a DNA sequences database of a protein sequence database, such as GenBank (E) and UniProt (F) [19, 20].

With the Bioinformatics, it is possible to analyze big data and to simulate complex systems that occur in nature, such as the functioning of a cell [15]. The development of researches in this area enables faster development and applications in the search for the treatment and prevention of various diseases. Thus, computer simulation relates data in order to generate a better response, having a better result in biology and health research.

Bioinformatics tools have been used to make inferences based on DNA sequences available in databases or generated from wet laboratory experiments. In these cases, program packages such as PreGap, Gap4 or 5 [16] and MEGA [17] are able to receive, edit and align sequences. In addition, organisms DNA sequences can be identified by local alignments using BLAST tool [18], which can be used to compare DNA or protein sequences available at public databases, as GenBank [19], UniProt [20], among others. These comparisons can be possible, for example, to identify genes proteins, and annotate their functions (Figure 1).

With Bioinformatics, it is possible to work with HPV using universal databases that provide sequences of various viruses, or databases specific for Papillomaviruses, such as the Papillomavirus Episteme (PaVE) [22]. Therefore, it can perform the comparison among sequences from different organisms, which allows the genotyping of HPV isolates (essential to differentiate high and low-risk HPV types).

In addition, Bioinformatics can also be used for modeling the 3D structure of HPV proteins, which permits the study of its function, structure and evolution in order to find therapeutic targets enabling the development of drugs for the treatment and prevention of papillomavirus lesions. Thus, many studies are carried out involving HPV and computational tools associated with structural genomics studies [23], phylogenetic analysis [24], epigenetics [25], protein 3D structure modelling [26, 27], among other activities.

BIOINFORMATICS TOOLS FOR TARGET IDENTIFICATION WITHIN DISEASE CONTROL

The usage of bioinformatics tools has arose significantly in the last decade, mainly due to its accuracy and capability to explore a wide range of molecular mechanisms in many diseases [28]. With the continuous advances in technology, computer sciences and molecular biology, it was possible for the bioinformaticians to develop new algorithms and, therefore, softwares focusing on disease control on a genomic level [28]. This allowed a more precise way of identifying key genes related to a certain disease [28, 29].

The use of RNA sequencing platforms has greatly contributed to the identification of differentially expressed genes, as well as the identification of large quantities of diagnostic or prognostic biomarkers of diseases such as cervical cancer [29, 30]. In general, biomarkers are molecules (i.e., proteins) present in body fluids or tissues, and these molecules can be used to monitor treatments, detect a disease in earlier stages and determine your prognostic [31, 32]. Currently, the only Food and Drug Administration (FDA) approved biomarkers for clinical use are the protein-based ones [33], and for this reason, we will henceforth focus on protein-based markers.

As new-generation sequencing techniques allow large amounts of biomarkers to be detected, protein microarrays have shown to be a robust tool in the search for new molecular targets for cancer, since it assesses protein-protein interactions, protein-ligand interactions and allow antibody profiling in various pathogenic conditions [34]. Protein of tissue microarrays have then set a milestone in proteomic studies, and although being a rather old kind of method, it is still useful for diagnostic and functional studies [33, 35]. The next paragraph will focus on differentially expressed gene data to reveal specific markers of cervical disease.

To select target genes and proteins as a biomarker, it is necessary to know the possible targets of interest very well. For this, the analysis of functional enrichment, metabolic pathway and a protein-protein interaction network (PPI) is indispensable. The performance of functional enrichment is important since it allows the gathering of information about a particular gene in terms of biological process, molecular function and cellular localization [36]. This analysis can be performed by searching the Gene Ontology (GO) (http://geneontology.org/) and Universal Protein Resource (UniProt) (https://www.uniprot.org/) databases. The second step in gathering information regarding the molecular interactions that the chosen gene performs, and for this, it is possible to use the Kyoto Encyclopedia of Genes and Genomes (KEGG) (https://www.genome.jp/kegg/) to map the metabolic pathways that its target participates. The third and final step to obtaining target information is knowing which are their partner genes and its interaction networks. Studies of protein-protein interactions have been conducted to identify host proteins that interact with pathogen proteins during the infection process. These studies are viewed as pioneers in understanding how the infectious process occurs and possible resistance mechanisms activated during infection, and also allows the survey of possible protein pathways for the development of therapeutic strategies and presents itself as a task of notorious interest in today's systems biology [37, 38].

There are currently several online databases where it is possible to search and verify the interaction of proteins of interest for a disease. Among the most common major public databases are Functional Protein Association (STRING) (https://string-db.org/), Interacting Proteins (DIP) (https://dip.doe-mbi.ucla.edu/dip/Main.cgi), Protein Interaction (IntAct) (https://www.ebi.ac.uk/intact/), Biological General Repository for Interaction Datasets (Bio GRID) (https://thebiogrid.org) and Molecular Interaction (MINT) (https://mint.bio.uniroma2.it/).

Basically, the very same techniques used for identifying targets in the host could be used for the pathogen, but the approach is slightly different in terms of treatment design. When analysing a disease like cancer for example, it is indispensable to take into account the transcription factors (TF). The reason for that is because the TFs, when malfunctioning, could act as oncogenes due to its control over multiple genes or tumour suppressors [30], which is the main cause of uncontrolled growth. Therefore, understanding the role of the TFs in gene regulation and its target genes is an important way to plan a possible treatment accordingly. For this task, the ChIP-seq technology, which stands for Chromatin Immunoprecipitation followed by Sequencing, may prove to be quite useful [39].

Along with ChIP-seq, other genome-scale sequencing technologies have risen in the past decade, namely DNA and RNA-seq. The latter two have very similar protocols, the main difference being the target molecule for analysis (e.g., DNA or RNA) [40, 41], while the first one relies on the association of transcription factors with DNA, as aforementioned. The RNA-seq guideline is based on the acquisition of a population of RNA, its conversion to a library of cDNA with adaptors attached to one or both ends, its sequencing in a high-throughput manner (single-end or pair-end sequencing) and its alignment to a reference genome [40, 41]. The result is a genome-scale transcriptional map that shows the level of expression for each gene [30, 40, 41].

Although serving different purposes, these techniques can be combined to map individual transcription factor regulons and entire regulatory networks, thus providing a deep comprehension of the disease mechanisms and enabling the search for new therapeutic targets [41].

In conclusion, biomarkers have great potential to act as therapeutic targets, monitor common molecular disease events such as in cervical cancer, improve detection of malignant lesions and identify the best therapeutic strategy, resulting in improvements in the prognosis of this disease.

Bioinformatics and Rational Drug Design

The development of new sequencing technologies and analysis tools provided several opportunities for understanding the molecular basis of human papillomavirus infection [42]. Through gene expression studies, it is possible to identify host genes and pathways that present altered regulation in the presence of the virus [43]. In addition, in recent years there has been an accumulation of information about the role of HPV proteins in the context of infection and cancer development [44].

The major viral oncoproteins (i.e., E6 and E7) interact with various host targets and are responsible for disrupting pathways associated with cell cycle control and apoptosis. The fact that the high expression of some HPV and human proteins are necessary to maintain the transformed status of infected cells opens possibilities for the rational design of inhibitors specific to these cells [45].

Despite recent technological advances, the time and cost for a new drug to enter the market is still very high. Estimates suggest that the price associated with developing a new drug can range from $92 to $883.6 million [46]. In this sense, Bioinformatics can be applied in different phases of drug design (e.g., identification, characterization and optimization of drug candidates) contributing to the whole process being more effective [47].

The pharmacological activity of a given substance is associated with the interaction of this molecule with specific sites present in the target, which results in the inhibition of processes involved in disease manifestation. Typically, to identify molecules with pharmacological activities of interest, pharmaceutical industries perform high-throughput screening (HTS) [48]. Based on this method, large libraries of molecules are tested in specific biological assays. Some limitations of HTS (e.g., high cost and time required) have been increasing the use of structure-based drug design (SBDD) in the routines of large pharmaceutical industries [48]. This method consists of using structural information to identify potential inhibitors of specific targets, usually using computational approaches.

A prerequisite for applying an SBDD approach is the target structure, also called the receptor. Commonly, this structure is obtained by experimental techniques such as X-ray crystallography and NMR. When the receptor structure is not available, in silico methods can be employed for the construction of 3D models. Some of the most commonly used protein modeling tools are Modeller [49], Swiss-Model [50] and I-Tasser [51].

Based on the target structure, several approaches can be used to identify inhibitor candidates. Among the options is the structure-based virtual screening (SBVS), considered the computational counterpart of the HTS method. In this approach, beyond the target structure libraries consisting of thousands or millions of small molecules are used, for example DrugBank

[52], ZINC [53] and PubChem [54]. SBVS consists of the prediction of ligand conformations relative to specific target sites and subsequent estimation of the binding energy of the complexes. Based on these analyzes, a subset of compounds with better binding affinities can be selected for further in vitro testing, rather than using the complete library. There are several public and commercial virtual screening tools. Some of the most used are AutoDock Vina [55], GOLD [56] and Glide [57].

Where the 3D structure of the receptor is not known, strategies based on knowledge of molecules that bind to targets or have targeted activities may be employed. This strategy is known as ligand-based drug design (LBDD). Quantitative structure–activity relationship (QSAR), a popular method of LBDD, is based on the premise that molecules with similar structures should exhibit similar activities [58]. In this method, the physicochemical or structural properties of the tested molecules are quantified, after the construction and application of the model, it is possible to predict new chemicals that potentially present the activity of interest [59, 60].

In the context of the development of new therapies against HPV, there is the possibility of applying the two mentioned approaches (i.e., SBDD and LBDD). Complete and partial structures of some virus proteins are available [61-64], enabling the use of structure-based approaches. Based on the structure of E6, a recent study conducted a virtual screening simulation and identified potential candidates for E6 inhibitors [27]. Another study used a structure-based approach to identify small molecules capable of blocking E6 oncoprotein function in vitro, resulting in decreased viability and proliferation of HPV-positive cells [65]. It should be noted that most studies in this field focus only on some genotypes, usually HPV16 and HPV18. As mentioned at the beginning of this chapter, over 200 types of HPV are currently known. In this sense, the inclusion of more genotypes in studies of this nature can be considered a challenge to be overcome.

Another point to consider is that there are no SBDD studies so far targeting host proteins. As mentioned at the beginning of this subtopic, inhibition of host proteins that are upregulated in the presence of the virus can be considered a promising strategy. In addition, once host targets that have a similar expression profile are identified in most HPV genotypes, it will be possible to develop nonspecific drugs that act against different viral types.

Regarding the application of ligand-based strategies, some studies have identified small molecules capable of disrupting HPV oncoprotein E6 functions, promoting senescence or apoptosis [65-67]. Based on the most active molecules, it is possible to perform QSAR studies to identify new molecules with similar profile. The other virus proteins (e.g., E7, E2 and E5) were poorly exploited so far.

As detailed above, for the identification of lead compounds that have activity against HPV-infected cells, studies should be developed that consider inhibition of both virus and host proteins. Using bioinformatics tools in parallel with specific bioassay routines can contribute to reducing the time and cost of developing new therapies against cervical cancer and other HPV-related diseases.

BIOINFORMATICS AND VACCINE DEVELOPMENT

Vaccines are pharmaceutical products that have good conditions (cost-benefit) at prevention or diseases treatment [68]. Attenuated antigens (usually dead or modified strains) that constitute microorganisms' structural elements are able to induce a host immune system response, generating an immunological memory that can be accessed if the pathogen comes into contact with the host [69].

Although vaccines are effective as a prophylactic strategy, they can also cause side effects, so in order to increase vaccine reliability, bioinformatics has been gaining ground in immunology studies, further improving the vaccine development process [70].

Bioinformatics is a research field that, in a multidisciplinary way, uses computational methods and information technology combined with biological interpretations aiming to produce and analyze molecular biology, genetics and biotechnology data [70].

Vaccine development requires a complete biological understanding about the disease causative agent. After obtaining pathogen data (in specialized database or literature), in silico predictions may be the way out for safer and more efficient vaccine-building compounds.

Some methods have been developed to improve vaccine production. One of those is Reverse vaccinology, a methodology that uses computational tools to identify structures that can induce immune response in the host, protecting it against specific diseases [68, 71]. This method has some advantages over the traditional method, since it reduces the time and costs in the vaccines development, besides increasing the number of proteins studied, which facilitates the antigen selection process [68].

According to [69], an interesting approach to the use of this method is the genomic information of the studied pathogen. In some cases, only with the pathogen sequence one can identify almost all proteins that can be expressed by using a bioinformatics software. Therefore, you just have to check which of these proteins have physicochemical properties that characterize them as good antigens (such as cell location, Adhesin properties, Antigenicity, Similarity, Transmembrane helix) [68].

Another possible approach is through immunoinformatics. The immune system can be divided into cellular or humoral, and depending on this classification, the responses may differ, so vaccine production must meet certain criteria [72, 73]. If the response is cellular, a good strategy is to locate antigens that can be recognized by major histocompatibility complex or MHC (proteins present in lymphocytes). Several software can perform this search, for example: TEpredict (http://tepredict.sourceforge.net/), CTLPred (http://crdd.osdd.net/raghava/ctlpred/), nHLAPred (http://crdd.osdd.net/raghava/nhlapred/), ProPred-I (http://crdd.osdd.net/raghava/propred1/), etc. In order to obtain data for this methodology, some specialized immunological databases can be consulted, such as the Free Epitope Database and Prediction Resource [74].

Another strategy is the structural vaccinology. In this approach, the focus is on the structural characteristics of the antigen candidate macromolecules. As seen in [68], structural vaccinology designs vaccines based on the selection or design of peptides or cross-reactive antigens, generating immunity against different antigens from divergent pathogens.

As seen in this chapter, computer simulations have been used to model pathogen-host interactions to improve vaccine development. In these processes, the distinction between immunogenic and non-immunogenic epitopes occurs, followed by simulation of immune response [70].

After evaluating of vaccine safety and efficacy, large databases of immunoinformatics and literature are added, providing support for the improvement of research already carried out. Several methods are being built or further refined to reproduce the behavior of the molecules analyzed as reliably as possible. Nevertheless, it is worth noting that existing methods offer vast and efficient resources for prospecting compounds or potential antigens for vaccines development.As seen in this chapter, computer simulations have been used to model pathogen-host interactions to improve vaccine development. In these processes, the distinction between immunogenic and non-immunogenic epitopes occurs, followed by simulation of immune response [70].

After evaluating of vaccine safety and efficacy, large databases of immunoinformatics literature are added, providing support for the improvement of research already carried out. Several methods are being built or further refined to reproduce the behavior of the molecules analyzed as reliably as possible. Nevertheless, it is worth noting that existing methods offer vast and efficient resources for prospecting compounds or potential antigens for vaccines development.

Conclusion

In conclusion, Bioinformatics tools have shown great potential to be used in studies that aimed at increasing knowledge on HPV infection pathways, developing more effective detection, and protection methods. The application of these and other computational methods may accelerate HPV-related discoveries, which is very important for cervical cancer control, a major public health problem worldwide. Therefore, in this context, novel studies focusing on the development or application of Bioinformatics tools for increasing HPV knowledge is paramount.

REFERENCES

[1] Hutter, J. N. and Decker, C. F. (2016). Human papillomavirus infection. *Disease-a-Month*, 62 (8): 294-300.

[2] Bray, F., Ferlay, J., Soerjomataram, I., Siegel, R. L., Torre, L. A. and Jemal, A. (2018). Global Cancer Statistics 2018: GLOBOCAN Estimates of Incidence and Mortality Worldwide for 36 Cancers in 185 Countries. *CA: A Cancer Journal for Clinicians*, 68: 394–424.

[3] Murillo, R. and Ordóñez-Reyes, C. (2019). Human papillomavirus (HPV) vaccination: from clinical studies to immunization programs. *International Journal of Gynecological Cancer*, doi: 10.1136/ijgc-2019-000582.

[4] Serrano, B., Alemany, L., Ruiz, P. A., Tousa, S., Lima, M. A., Bruni, L., Jain, A., Clifford, G. M., Qiao, Y. L., Weiss, T., Bosch, X. F. and Sanjose, S. (2014). Potential impact of a 9-valent HPV vaccine in HPV-related cervical disease in 4 emerging countries (Brazil, Mexico, India and China). *Cancer Epidemiology*, 38: 748–756.

[5] Handler, M. Z., Handler, N. S., Majewski, S. and Schwartz, R. A. (2015). Human papillomavirus vaccine trials and tribulations: Clinical perspectives. *Journal of the American Academy of Dermatology*, 73: 743-756.

[6] Briantini, P., Flammineis, E. and Mercuri, S. R. (2017). Review of HPV-related diseases and cancers. *New Microbiologica*, 40 (2): 80-85.

[7] Palefski, J. M. (2016). Epidemiology of human papillomavirus infections. In: *UpToDate, Post TW (Ed)*, UpToDate, Waltham, MA. (Acessed on August 29).

[8] Groves, I. J. and Coleman, N. (2015). Pathogenesis of human papillomavirus-associated mucosal disease. *The Journal of Pathology*, 235: 527–538.

[9] Egawa, N., Egawa, K., Griffin, H. and Doorbar, J. (2015). Human papillomaviruses; epithelial tropisms, and the development of neoplasia. *Viruses*, 7: 3863–3890.

[10] Negahdaripour, M., Eslami, M., Nezafat, N., Hajighahramani, N., Ghosshon, M. B., Shoolian, E., Dehshahri, A., Erfani, N., Morowvat, M. H. and Ghasemi, Y. (2017). A novel HPV prophylactic peptide vaccine, designed by immunoinformatics and structural vaccinology approaches. *Infection, Genetics and Evolution*, 54: 402-416.

[11] Amorim, A. T., Marques, L. M., Campos, G. B., Lobão, T. N., Lina, V. S., Cintra, R. C., Andreoli, M. A., Villa, L. L., Boccardo, E., Braga Júnior, A. C. R., López, R. V. M., Santos, D. B., Souza, G. M., Romano, C. C. and Timenetsky, J. (2017). Co-infection of sexually transmitted pathogens and Human Papillomavirus in cervical samples of women in Brazil. *BMC Infectious Diseases*, 17: 769.

[12] GLOBOCAN. (2012). *Estimated cancer incidence, mortality and prevalence worldwide*. (Accessed on August 29). https://www.iarc.fr/news-events/latest-world-cancer-statistics-globocan-2012-estimated-cancer-incidence-mortality-and-prevalence-worldwide-in-2012/.

[13] Tomen, E. A., Yohanna, S. and Obilom, R. E. (2019). Awareness of Papanicolaou smear and visual inspection with acetic acid as screening tools for cervical cancer among women attending the General Outpatient Clinic of a tertiary institution in north central Nigeria. *Nigerian Medical Journal*, 60 (2): 76-79.

[14] Eo, Y. S. and Kim, J. S. (2019). Associations of health belief and health literacy with Pap smear practice among Asian immigrant women. *European Journal of Oncology Nursing*, 42: 63-68.

[15] Luscombe, N. M., Greenbaum, D. and Gerstein, M. (2001). What is Bioinformatics? A Proposed Definition and Overview of the Field. *Methods of Information in Medicine*, 40: 346–58.

[16] Staden, R. (1996). The Staden Sequence Analysis Package. *Molecular Biotechnology*, 5: 233-241.

[17] Tamura, K., Stecher, G., Peterson, D., Filipski, A. and Kumar, S. (2013). MEGA6: Molecular Evolutionary Genetics Analysis Version 6.0. *Molecular Biology and Evolution*, 30: 2725-2729.

[18] Altschul, S. F., Gish, W., Miller, W., Myers, E. W. and Lipman, D. J. (1990). Basic local alignment search tool. *Journal of Molecular Biology*, 215: 403-410.

[19] GenBank. (2019). *NIH genetic sequence database*. Accessed September 2019. https://www.ncbi.nlm.nih.gov/genbank/.

[20] UniProt. (2019). *Universal Protein Resource*. Accessed September 2019. https://www.uniprot.org/.

[21] Barbosa Filho, R. A. A. (2011). *Cloning, sequencing and molecular studies of the isolated HPV 16 genome in the Amazon*. PhD diss., Federal University of Amazonas.

[22] PAVE. (2019). *Papillomavirus Episteme*. Accessed September 2019. https://pave.niaid.nih.gov/.

[23] Akagi, K., Li, J., Broutian, T. R., Padilla-Nash, H., Xiao, W., Jiang, B., Rocco, J. W., Teknos, T. N., Kumar, B., Wangsa, D., He, D., Ried, T., Symer, D. E. and Gillison, M. L. (2014). Genome-wide analysis of HPV integration in human cancers reveals recurrent, focal genomic instability. *Genome Research.* 24 (2): 185-99.

[24] Bogaards, J. A., van der Weele, P., Woestenberg, P. J., van Benthem, B. H. B. and King. A. J. (2019). Bivalent Human Papillomavirus (HPV) Vaccine Effectiveness Correlates With Phylogenetic Distance From HPV Vaccine Types 16 and 18. *The Journal of Infectious Diseases*, 220: 1141–1146.

[25] Yuan, M., Yuan, J., Mei, L. and Abulizi, G. (2018). Bioinformatics analysis of methylation in cervical adenocarcinoma in Xinjiang, China. *Medicine*, 97 (35): 1-6.

[26] Kumar, A., Rathi, E. and Kini, S. G. (2019). E-pharmacophore modelling, virtual screening, molecular dynamics simulations and in-silico ADME analysis for identification of potential E6 inhibitors against cervical cancer. *Journal of Molecular Structure*, 1189: 299-306.

[27] Ricci-López, J., Vidal-Limon, A., Zunñiga, M., Jimènez, V. A., Alderete, J. B., Brizuela, C. A. and Aguila, S. (2001). Molecular modeling simulation studies reveal new potential inhibitors against HPV E6 protein. *Plos One*, 14: 1-22.

[28] Cheng, C., Hua, J., Tan, J., Qian, W., Zhang, L. and Hou, X. (2019). Identification of differentially expressed genes, associated functional terms pathways, and candidate diagnostics biomarkers in inflammatory bowel diseases by bioinformatics analysis. *Experimental and Therapeutical Medicine*, 18: 278-288.

[29] Essebier, A., Lamprecht, M., Piper, M. and Bodén, M. (2017). Bioinformatics approaches to predict target genes from transcription factor binding data. *Methods*, 131: 111-119.

[30] Bhagwat, A. S. and Vakoc, C. R. (2015). Targeting transcription factors in cancer. *Trends in Cancer*, 1 (1): 53-65.

[31] Hu, B., Niu, X., Cheng, L., Yang, L. N., Li, Q., Wang, Y., Tao, S. C. and Zhou, S. M. (2015). Discovering cancer biomarkers from clinical samples by protein microarrays. *Proteomics Clinical Applications*, 9 (1-2): 98-110.

[32] Olszewska-Słonina, D. and Czerniak, B. (2018). Biomarkers could facilitate prediction of worse clinical outcome of cancer with special insight to cervical cancer. *Contemporary Oncology*, 22 (1): 1-7.

[33] Lim, S. Y., Lee, J. H., Diefenbach, R. J., Kefford, R. F. and Rizos, H. (2018). Liquid biomarkers in melanoma: detection and Discovery. *Molecular Cancer*, 17 (8): 1-14.

[34] Atak, A., Mukherjee, S., Jain, R., Gupta, S., Singh, V. A., Gahoi, N., Manubhai, K. P. and Srivastava, S. (2016). Protein microarrays applications: autoantibody detection and posttranslational modification. *Proteomics*, 16 (19): 2557-2569.

[35] Perez-rivas, L. G., Ierez, J. M., Sousa, C. E. F., Luque, V., Quero, C., Pajares, B. Franco, L., Sanchez-muñoz, A., Ribelles, N. and Alba, E. (2012). Serum protein levels following surgery in breast cancer patients: A protein microarray approach. *International Journal of Oncology*, 41 (6): 2200-2206.

[36] Berardini, T. Z., Mundodi, S., Reiser, L., Huala, E., Garcia-Hernandez, M., Zhang, P., Mueller, L. A., Yoon, J., Doyle, A., Lander, G., Moseyko, N., Yoo, D., Xu, I., Zoeckler, B., Montoya, M., Miller, N., Weems, D. and Rhee, S. Y. (2004). Functional annotation of the

Arabidopsis genome using controlled vocabularies. *Plant Physiology*, 135 (2): 745-55.

[37] Zhang, C. Q., Petrey, D., Garzo, J. I., Deng, L. and Honig, B. (2011). PrePPI: a structure-informed database of protein–protein interactions. *Nucleic Acids Research*, 41. doi: 10.1093/nar/gks1231.

[38] Feinauer, C., Szurmant, H., Weigt, M. and Pagnani, A. (2016). Inter-Protein Sequence Co-Evolution Predicts Known Physical Interactions in Bacterial Ribosomes and the Trp Operon. *Plos One*, 11 (1): 1-18.

[39] Wade, J. T. (2015). Mapping transcriptional regulatory networks with ChIP-seq and RNA-seq. *Advances in Experimental Medicine and Biology*, 833: 119-134.

[40] Wang, Z., Gerstein, M. and Snyder, M. (2009). RNA-Seq: a revolutionary tool for transcriptomics. *Nature Reviews Genetics*, 10 (1): 57-63.

[41] Klymenko, T., Gu, Q., Herbert, I., Stevenson, A., Iliev, V., Watkins, G., Pollock, C., Bhatia, R., Cuschieri, K., Herzyk, P., Colector, D. and Graham, S. V. (2012). RNA-Seq analysis of differentiated keratinocytes reveals a massive response to late events during human papillomavirus 16 infection, including loss of epithelial barrier function. *Journal of Virology*, doi: 10.1128/JVI.01001-17.

[42] Yang, R., Klimentová, J., Göckel-krzikalla, E., Ly, R., Gmelin, N., Hotz-wagenblatt, A., Rehulkova, H., Stulik, J., Rosl, F. and Niebler, M. (2019). Combined Transcriptome and Proteome Analysis of Factors and Networks in HPV-Induced Carcinogenesis. *mSphere*, 4 (2): 1–23.

[43] Schmitt, M. and Pawlita, M. (2011). The HPV transcriptome in HPV16 positive cell lines. *Molecular and Cellular Probes*, 25 (2–3): 108–113.

[44] Tomaić, V. (2016). Functional roles of E6 and E7 oncoproteins in HPV-induced malignancies at diverse anatomical sites. *Cancers*, 8 (10). doi: 10.3390/cancers8100095.

[45] Suarez, I. and Trave, G. (2018). Structural insights in multifunctional papillomavirus oncoproteins. *Viruses*, 10 (1): 1–22.

[46] Morgan, S., Grootendorst, P., Lexchin, J., Cunningham, C. and Greyson, D. (2011). The cost of drug development: A systematic review. *Health Policy*, 100 (1): 4–17.

[47] Ekins, S. (2006). *Computer applications in pharmaceutical research and development.* John Wiley & Sons; Hoboken, NJ: 2006.

[48] Lionta, E., Spyrou, G., Vassilatis, D. K. and Cournia, Z. (2014). Structure-Based Virtual Screening for Drug Discovery: Principles, Applications and Recent Advances. *Current Topics in Medicinal Chemistry*, 14: 1923–1938.

[49] Eswar, N., Webb, B., Marti-renom, M. A., Madhusudhan, M. S., Eramian, D., Shen, M., Pieper, U. and Sali, A. (2006). Comparative protein structure modeling using Modeller. *Current Protocols in Bioinformatics*, doi: 10.1002/0471250953.

[50] Biasini, M., Bienert, S., Waterhouse, A., Arnold K., Studer G., Schmidt, T., Kiefer, F., et al. (2014). SWISS-MODEL: Modelling protein tertiary and quaternary structure using evolutionary information. *Nucleic Acids Research,* 42 (W1): 252–258.

[51] Yang, J., Yan, R., Roy, A., Xu, D., Poisson, J. and Zhang, Y. (2014). The I-TASSER suite: Protein structure and function prediction. *Nature Methods*, 12 (1): 7–8.

[52] Wishart, D., Knox, S. C., Guo, A. C., Cheng, D., Shrivastava, S., Tzur, D., Gautam, B. and Hassanali, M. (2008). DrugBank: A knowledgebase for drugs, drug actions and drug targets. *Nucleic Acids Research*, 36 (SUPPL. 1): 901–906.

[53] Irwin, J. J. and Shoichet., B. K. (2005). ZINC - A free database of commercially available compounds for virtual screening. *Journal of Chemical Information and Modeling*, 45 (1): 177–182.

[54] Kim, S., Thiessen, A. P., Bolton, E. E., Chen, J., Fu, G., Gindulyte, A., Han, L., et al. (2016). PubChem substance and compound databases. *Nucleic Acids Research*, 44 (D1): D1202–D1213.

[55] Trott, O. and Olson, A. (2010). Autodock vina: improving the speed and accuracy of docking. *Journal of Computational Chemistry*, 31 (2): 455–461.

[56] Jones, G., Willett, P., Glen, R. C., Leach, A. R. and Taylor, R. (1997). Development and validation of a genetic algorithm for flexible docking. *Journal of Molecular Biology*, 267(3), 727–748.

[57] Friesner, R. A., Banks, J. L. and Murphy, R. B. (2004). Glide: a new approach for rapid, accurate docking and scoring. *Journal of Medicinal Chemistry*, 47 (7): 1739–1749.

[58] Acharya, C., Coop, A., E. Polli, J. and D. MacKerell, A. (2011). Recent Advances in Ligand-Based Drug Design: Relevance and Utility of the Conformationally Sampled Pharmacophore Approach. *Current Computer Aided-Drug Design*, 7(1), 10–22.

[59] Kurogi, Y. and Guner, O. F. (2001). Pharmacophore modeling and three-dimensional database searching for drug design using catalyst. *Current Medicinal Chemistry*, 8 (9): 1035–1055.

[60] Mason, J., Good, A. and Martin, E. (2005). 3-D Pharmacophores in Drug Discovery. *Current Pharmaceutical Design*, 7 (7): 567–597.

[61] Zanier, K., Charbonnier, S., Sidi, A. O. M. O., McEwen, A. G., Ferrario, M. G., Poussin-Courmontagne, P., et al. (2013). Structural Basis for Hijacking of Cellular LxxLL Motifs by Papillomavirus E6 Oncoproteins. *Science*, 339 (6120): 694–698.

[62] Liu, X., Clements, A., Zhao, K. and Marmorstein, R. (2006). Structure of the human Papillomavirus E7 oncoprotein and its mechanism for inactivation of the retinoblastoma tumor suppressor. *Journal of Biological Chemistry*, 281 (1): 578–86.

[63] Ohlenschläger, O., Seiboth, T., Zengerling, H., Briese, L., Marchanka, A., Ramachandran, R., Baum, M., et al. (2006). Solution structure of the partially folded high-risk human papilloma virus 45 oncoprotein E7. *Oncogene*, 25 (44): 5953–5959.

[64] Abbate, E. A., Voitenleitner, C. and Botchan, M. R. (2006). Structure of the Papillomavirus DNA-Tethering Complex E2:Brd4 and a Peptide that Ablates HPV Chromosomal Association. *Molecular Cell*, 24 (6): 877–889.

[65] Cherry, J. J., Rietz A., Malinkevich, A., Liu, Y., Xie, M., Bartolowits, M., Davisson, V. J., Baleja, J. D. and Androphy, E. J. (2013). Structure

based identification and characterization of flavonoids that disrupt human papillomavirus-16 E6 function. *PLoS One*, 8 (12): 1–21.

[66] Baleja, J. D., Cherry, J. J., Liu, Z., Gao, H., Nicklaus, M.C., Voigt, J. H., Chen, J. J. and Androphy, E. J. (2006). Identification of inhibitors to papillomavirus type 16 E6 protein based on three-dimensional structures of interacting proteins. *Antiviral Research*, 72 (1): 49–59.

[67] Malecka, K., Fera, D. A., Schultz, D. C., Hodawadekar, S., Reichman, M., Donover, P. S., Murphy, M.E. and Marmorstein, R. (2014). Identification and characterization of small molecule human papillomavirus E6 inhibitors. *ACS Chemical Biology*, 9 (7): 1603–1612.

[68] Ribas-Aparicio, M. R., Arturo, C. J., Alicia, J. A., Paulina, M. G. and Gerardo, A. O. (2017). The Impact of Bioinformatics on Vaccine Design and Development. *Vaccines,* doi:10.5772/intechopen.69273.

[69] Jarzab, A., Skowicki, M. and Witkowska, D. (2013). Subunit Vaccines – Antigens, Carriers, Conjugation Methods and the Role of Adjuvants. *Postępy Higieny i Medycyny Doświadczalnej*, 67: 1128–1143.

[70] Soria-Guerra, R. E., Nieto-Gomez, R., Govea-Alonso, D. O. and Rosales-Mendoza, S. (2014). An Overview of Bioinformatics Tools for Epitope Prediction: Implications on Vaccine Development. *Journal of Biomedical Informatics*. Elsevier Inc. doi:10.1016/j.jbi.2014.11.003.

[71] Monterrubio-López, G. P., González-Y-Merchand, J. A. and Ribas-Aparicio, M. R. (2015). Identification of Novel Potential Vaccine Candidates against Tuberculosis Based on Reverse Vaccinology. *BioMed Research International*, 2015: 11–14.

[72] Pandey, K., Sharma, M., Saarav, I., Singh, S., Dutta, P., Bhardwaj, A. and Sharma, S. (2016). Analysis of the DosR Regulon Genes to Select Cytotoxic T Lymphocyte Epitope Specific Vaccine Candidates Using a Reverse Vaccinology Approach. *International Journal of Mycobacteriology*, 5 (1): 34–43.

[73] Negar, S., Taheri, T. and Rafati, S. (2016). Post-Genomics and Vaccine Improvement for Leishmania. *Frontiers in Microbiology*, 7: 1–13.

[74] National Institutes of Health in the Department of Health and Human Services. (2019). *Immune Epitope Database*. (Accessed August 25). https://www.iedb.org/home_v3.php.

In: Human Papillomaviruses (HPVs)
Editor: Yvette Franklin
ISBN: 978-1-53617-331-4

Chapter 3

PROPHYLAXIS OF CANCERS FROM HUMAN PAPILLOMA VIRUSES (HPV'S)

Louis Z. G. Touyz*, BDS; MSc(Dent); MDent (Perio & Oral Med)*
McGill University, Faculty of Dentistry
Montreal, PQ, Canada

ABSTRACT

HPV [Human Papilloma Virus] is one known cause for inducing oncogenesis. This chapter appraises HPV vaccines for prophylaxis of HPV infection in healthy individuals. Effective vaccines exist to ensure prophylaxis against HIV morbidity and mortality. Inoculation against HPV, to acquire immunity, is advised for all individuals before sexual debut.

* Corresponding Author's Email: louis.touyz@mcgill.ca; Touyzlouis@gmail.com.

Introduction

Etiologies and pathogenesis of cancer is complex and only partially understood. Some causes of selected cancers have been defined: Asbestos induces mesothelioma, radiation induces neoplasias, aniline dyes cause urogenital carcinogenesis, nitrosamines precipitate oncogenesis in the Gastro-intestinal tract, cyclic petrochemicals promote carcinogens, genes (*BRCA,* for instance) predispose to certain types of breast cancer, smoking or chewing tobacco generates lung or oral cancer, and ethanol stimulates neoplastic change [1]. The human papilloma virus (HPV) was deemed benign for decades, with few variants [2-4]. But biomedical research revealed 150 serologically identifiable varieties, some benign, but many with oncogenic genes [4, 5]. To boost humoral responses to viral antigens, immune anamnestic responses are stimulated to prevent disease and most subsequent recrudescence [6-8]. Vaccines against some of the earlier varieties are Gardasil (Merck, Kenilworth, NJ, U.S.A.) and Cervarix (GlaxoSmithKline, Brentford, U.K.), was introduced and targeted at girls before their sexual debut [9]. Some religious authorities railed against HPV vaccination, claiming HPV vaccination promotes promiscuity [10]. To reduce societal sources of residual infection, HPV vaccination advice included boys [9].

Most vaccines successfully impart life-long immunity. Consequently, the earlier the vaccination, the longer and stronger the likelihood of preventing HPV morbidity [11].

Aim: This chapter appraises HPV vaccines for prophylaxis of HPV infection in healthy individuals.

Protection and New Vaccine

Initial immunity to HPV oncogenesis was realized from 4 HPV varieties; Gardasil, a quadrivalent vaccine, protects against HPV-16, HPV-18, HPV-6, and HPV-11. The additional HPVs [6 and 11] included in the Gardasil formulation are causally related to genital condylomata and recurrent respiratory papillomatosis. Other HPV's involved in preventing oncogenesis are: HPV-31, HPV-33, HPV-45 HPV-52, and HPV-58 [12-14]. These HPV vaccines were successfully used in national vaccination campaigns [15]. Because immunity is rarely completely induced with one vaccinating inoculation, there is a small but significant group who initially fail to respond adequately, and require triple inoculations to acquire immunity [12-14]. Usually the HPV-vaccine is administered as an intra-muscular injection, the first shot being followed by a second dose 2 months after the first dose, and the third shot 6 months after the first dose [16]. The initial protection from early vaccines had limits, and although high-levels of anti–HPV16 and HPV18 antibodies persist for up to 7 years, the triple boosters are recommended. Subsequently a broader-based HPV vaccine called V503 (with activity against 9 HPV varieties) was produced. In 2014, the bio-pharmaceutical firm Merck, successfully made a Biologics License Application for registration of the product, which was approved by the U.S. Food and Drug Administration [16].

Discussion

Despite the characterization of many causes of cancer, absolute cures remain obstinately obscure. Prophylaxis is still the most desirable approach, and for cause-defined cancers, prevention with immunity is partially, but significantly, successful. Chemotherapy, resective surgery and radiation were the major foundation therapies for decades to manage established cancer, but they had limited success, with notable morbidity and mortality. Now (2019) the HPV vaccinations stimulate immunomodulation and contribute to the prevention and lowering the prevalence of HPV-positive urogenital cancers, head-and-neck squamous cell carcinomas and recurrent respiratory papillomatosis [3, 17]. Consequently inoculation against HPV should be mandatory for all girls and boys prior to their sexual debut [9, 10]. Transmission of HPV can occur with genital or oro-genital contact and often with social kissing or droplet spread through inhalation from respiratory sprays of coughs or sneezes from hpv carriers [18, 19]. In the mid-1950s contraceptive hormone pills were introduced, barrier methods of contraception diminished, and sexual behaviors and mores changed; subsequently HPV, together with other sexually transmitted diseases, became more prevalent. Countries which were slow to adopt vaccination against HPV are prone to increased HPV morbidity and mortality [20, 21]. When people are infected with Human Immunodeficient Virus [HIV], they are susceptible to HPV infection and HPV morbidity [22]. Serious counselling and inoculation in HIV+ve females is required to avoid morbidity before AIDS materializes [23].

Vaccination protection takes a long time to wane and can last for decades [8, 14]. HPV viruses are implicated in cancers other than anogenital neoplasias [23-25]. Vaccination prophylaxis is effective against over 90% of HPV induced cancers and should be considered essential for all and become part of the essential inoculations administered in the first decade of life [12, 15].

Should HPV-vaccination be avoided a HPV pandemic is predicted, and the use of broad-based vaccines is desirable to optimize benefits from HPV prophylaxis [25]. Both sexes if not vaccinated will contract HPV infections [28, 29]. The broad based vaccines protect against HPV types 16, 18, 31, 33, 45, 52 and 58; vulvar and vaginal cancers are caused by the same HPV types. Genital warts, oral condylomata and recurrent respiratory papillomatosis are caused by HPV types 6 and11 [17]. Vaccination against HPV also boosts immunity against other viral infections such as mumps [26]. Proper inoculations with Gardasil®9 produces immunity against these 9HPV's and prevents oncogenesis by inducing effective immunity which lasts for decades [27].

Infections with single types of HPV (like HPV 52, 31, 33, 35, 39, 45, 51, 56, or 59) are not as oncogenic as simultaneous coinfections concomitant with type HPV 16. The four most dominant HPV genotypes found in Chinese women are HPV 16, 52, 58, 18. Patients with a single HPV 16 infection had higher incidences of CIN2+. [CIN-Cervical Intraepithelial Neoplasia Grade1 -> 5) Patients coinfected with HPV 16 had higher CIN2+ incidence than single HPV infections [28].

Non-vaccinated men also will be infected with HPV. Heterosexual and homosexual behaviors are exposed to HPV, with heterosexuals being more prevalent. HPV 32, 18, 16 and 58 were frequently identified and HPV 11, 6,81 and 61 are the most frequently low risk HPVs observed in a cohort of Chinese men [29].

Worldwide yearly cancer incidence is 14 million new cases, with a resultant death of 8 million individuals. Prevention of this mortality is the main mission. HPV inoculation can significantly reduce this frequency. In delivery of health care, anamnestic or freshly acquired immunity is used as prophylaxis against recrudescent cancer when patients are in remission [29].

Rationally vaccines are given to healthy individuals to induce immunity. Vaccinations may be prophylactic or therapeutic. ImMucin (Vaxil Israel, Rehovot, Israel) is administered to people who already infected, and yet the vaccine acts like a drug with physiological effects, behaving as an agent between a therapeutic drug and a vaccine, with a two-pronged hybrid therapy. Attacking target-viruses predisposing to cancer is what vaccines accomplish; they induce an immune reaction to neutralize viruses implicated in neoplasias. Vaxil claims their induced immunity targets mutated cells with early oncological changes, and consequently lowers the incidence of early cancers without affecting healthy cells. Vaxil Israel reports success with multiple myeloma and has tried the same approach on other cancers [29, 30].

p53 protein is a moderator of oncogenesis as a tumor suppressor. Protein 53 is activated by a variety of cellular signals, including DNA damage. P53 activates caspases which are the effectors of apoptosis. Induced neoplasia halts apoptosis and promotes uncontrolled cell growth without natural cell death. Some viruses block activation of apoptosis by blocking the caspase-mediated pathway. Lesions at the p53 locus are among the most frequently occurring mutations in human cancer. Loss or inactivation of p53 function strongly predisposes a cell to accumulate additional mutations that may further its progression to neoplastic change. The HPV protein p53 can be detected during histopathology examination, and a search for proteins to moderate protein p53 is needed [31].

Conclusion

Clinical tests to detect HPV infection are available, and vaccines will protect against HPV morbidity. All children prior to their sexual debut should be inoculated against HPV. Also because of the proven immune response, all adults who are not inoculated against HPV, are sexually active, and face the probability of more than one sexual partner, are well advised to be vaccinated against hpv. Further research into the clinical management of hpv is needed.

References

[1] American Cancer Society (acs). *What causes cancer?* [Web page]. Atlanta, GA: acs; (2016) [Available online at: http://www.cancer.org/cancer/cancercauses; cited 25 October 2015]

[2] Jensen AB, Link CC, Lancaster WD. Section ii. Papilliomavirus etiology of oral cavity papillomas. In: Hooks JJ, Jordan GW, eds. *Viral Infections in Oral Medicine*. Oxford, U.K.: Elsevier Science and Technology; 1982: 132–46.

[3] Norkin LC.(2010) Virology: *Molecular Biology and Pathogenesis*. Washington, DC: American Society for Microbiology Press; 419–43.

[4] Lajer CB. (2010) von Buchwald C. The role of human papillomavirus in head and neck cancer. *APMIS.*118:510–19.

[5] Touyz LZG. (2013) Human papilloma virus (HPV)—a biological and clinical appraisal: 2013. *Science Postprint*. 1:e00001.

[6] Forman D, de Martel C, Lacey CJ, et al., (2012) Global burden of human papillomavirus and related diseases. *Vaccine* 2012;30 (suppl 5): F12–23.

[7] Crosignani P, De Stefani A, Fara GM, et al., (2013)Towards the eradication of hpv infection through universal specific vaccination. *BMC Public Health.*13:642.

[8] Schiller JT, Castellsagué X, Garland SM. (2012) A review of clinical trials of human papillomavirus prophylactic vaccines. *Vaccine.* 30 (suppl 5): F123–38.

[9] Malagón T, Drolet M, Boily MC, et al., (2012)Cross-protective efficacy of two human papillomavirus vaccines: a systematic review and meta-analysis. *Lancet Infect. Dis.* 12: 781–9.

[10] Touyz LZG. (2009) Human papilloma virus inoculation: why only girls? *Curr. Oncol*. 16:2.

[11] Touyz SJJ, Touyz LZG.(2013) The kiss of death: hpv rejected by religion. *Curr. Oncol*. 20: e52–3.

[12] Touyz LZG (2016) HPV Vaccines; Still needed. *Current Oncology*. 23,4;e330-331.

[13] Schiller JT, Castellsagué X, Garland SM. (2012) A review of clinical trials of human papillomavirus prophylactic vaccines. *Vaccine.* 30(suppl 5): F123–38.

[14] Malagón T, Drolet M, Boily MC, et al., (2012) Cross-protective efficacy of two human papillomavirus vaccines: a systematic review and meta-analysis. *Lancet Infect. Dis.* 12: 781–9.

[15] Safaeian M, Porras C, Pan Y, et al., (2016) on behalf of the cvt group. Durable antibody responses following one dose of the bivalent human papillomavirus L1 virus-like particle vaccine in the Costa Rica Vaccine Trial. *Cancer Prev. Res*. (Phila);6: 1242–50.

[16] Liu X, Feng A, Cui Y, Tobe RG. (2013) Prevention of human papillomavirus (hpv) infection and cervical cancer in China: how does hpv vaccination bring about benefits to Chinese women? *Biosci. Trends.* 7:159–67.

[17] Merck and Co. (2014) *FDA Accepts for Review Merck's Biologics License Application for V503, Investigational 9-Valent Human Papillomavirus Vaccine* . Whitehouse Station, NJ: Merck, Sharp and Dohme Corp.; [Available online at: http://www.mercknewsroom.com/news-release/research-and-development-news/fda-accepts-review-mercks-biologics-license-application-v; cited 14 June 2016] Drug Identification number: DIN For Gardasil®9 -02437085.

[18] Syrjänen S. (2010) Current concepts on human papilloma virus infections in children. *APMIS*. 118:494–509.

[19] Chaturvedi AK, Engels EA, Pfeiffer RM, et al., (2011) Human papillomavirus and rising oropharyngeal cancer incidence in the United States. *J. Clin. Oncol.* 29:4294–301.

[20] Touyz LZG. (2009) Lips, kissing and oral implications. *Jnl. Aesthetic Dentistry*, 29–34.

[21] Nasman A, Attner P, Hammerstedt L, et al., (2009) Incidence of human papilloma virus (hpv) positive tonsillar carcinoma in Stockholm, Sweden: an epidemic of viral-induced carcinoma? *Int. J. Cancer.* 125:362–6.

[22] Bender E. (2014) Developing world: global warning. Sci Am 2014;311:S18–19. [Republished as: Bender E. Developing world: global warning. *Nature*. 509:S64–5].

[23] Chaturvedi AK, Madeleine MM, Biggar RJ, Engels EA. (2009)Risk of human papillomavirus-associated (HPV)cancers among persons with aids. *J. Natl. Cancer Inst.* 101(16):1120–30.

[24] Kenya S, Carrasquillo O, Fatil M, et al., (2015) Human papilloma virus and cervical cancer education needs among hiv-positive Haitian women in Miami. *Women Health Issues.* 25(3):262–6.

[25] Heng B, Glenn WK, Ye Y, et al., (2009) Human papilloma virus is associated with breast cancer. *Br. J. Cancer.* 101:1345–50.

[26] Massad LS, Einstein MH, Huh WK, et al., (2013) On behalf of the 2012 asccp Consensus Guidelines Conference. 2012 Updated consensus guidelines for the management of abnormal cervical cancer screening tests and cancer precursors. *Obstet. Gynecol.* 121:829–46.

[27] Pashley NR. (2002) Can mumps vaccine induce remission in recurrent respiratory papilloma? *Arch. Otolaryngol. Head Neck Surg.* 128:783–6.

[28] Kojic EM, Kang M, Cespedes MS, et al., (2014) Immunogenicity and safety of the quadrivalent human papillomavirus vaccine in hiv-1-infected women. *Clin. Infect. Dis*. 59(1):127–35.

[29] Li M, Du X, Lu M, Zhang W, Sun Z, Li L. et al., (2019) Prevalence characteristics of single and multiple HPV infections in women with

cervical cancer and precancerous lesions in Beijing China. *Journal Medical Virology*. Oct 2019.

[30] Xin HN, LI HJ, Li Z, LiXW. (2017) Genital infection among heterosexual and homosexual male attendees of sexually transmitted diseases clinic in Beijing China. *Epidemiology and Infecetio*. Aug. 145;13: 2838-2847.

[31] Neff J. (2016) New Israeli Cancer Vaccine Triggers Response in 90% of Cancer Types [Web article]. Herzliya, Israel: *NoCamels.com*; 2015. [Available at: http://nocamels.com/2015/01/new-vaccine-for-cancer/; 14 June 2016].

[32] p53 *Cell-Based Assay* (2015) Contact Signosis, Inc.1700 Wyatt Dr Suite 10; Santa Clara, California | 95054. Tel. 1–877–266–8388. [Available online at: www.signosisinc.com; cited dd Month year].

Authors' statement: The author has no conflict of interest to declare.

BIBLIOGRAPHY

A manual for cervical cancer screening and control: principles, practice, and new perspectives

LCCN	2013933799
Type of material	Book
Personal name	Branca, Margherita.
Main title	A manual for cervical cancer screening and control: principles, practice, and new perspectives / Margherita Branca and Adhemar Longatto Filho.
Published/Produced	New York: Nova Science Publishers, Inc., [2014] ©2014
Description	ix, 311 pages: illustratrions (some color): 25 cm.
ISBN	9781634630108 1634630106
LC classification	RC280.U8 B69 2014
Related names	Filho, Adhemar Longatto. Branca, Margherita edt
Summary	Cervical cancer (CC) is a major global health problem. Worldwide, an estimated 530,000 new cases of CC and 275,000 deaths from this disease are encountered every year. CC affects

particularly women less than 45 years of age and it is the leading cause of cancer-related deaths among women in the developing and underdeveloped countries, where 60-80% of cases are detected in advanced stages. After Human Papillomavirus (HPV) identification as the infectious etiological agent and elucidation of the natural history of CC, this tumor has become the most preventable and curable of all cancers. Well-organized screening programs with Papanicolaou (Pap) smear are known to be successful in decreasing both mortality and morbidity of this disease. This book describes the optional screening programs for CC prevention, whether organized, opportunistic or spontaneous and also their merits and weaknesses. Monitoring of the pathways and procedures of CC screening are mandatory in order to minimise the risks (wrong diagnosis, over-diagnosis, over-treatment) and maximize the benefits. In an optimal setting, the women should be the focus of the screening intervention and receive basic information on the crucial points of CC screening, so as to improve women's collaboration and compliance. The availability of alternative new screening tools to detect pre-cancer lesions provide new opportunities for CC prevention, as do HPV vaccines implemented in early life before viral exposure. This book, rich in pictures, tables and flowcharts is a comprehensive treatise on CC screening: its 14 chapters illustrate the fundamental principles and the necessary interventions for CC prevention and control and a guide for all medical and paramedical

	professionals involved in the interplay of CC screening.-- Source other than Library of Congress.
Contents	Chapter 1. The global burden of cervical cancer -- Chapter 2. Opportunities for cervical cancer prevention: the primary prevention -- Chapter 3. Human Papillomavirus (HPV) infections -- Chapter 4. Risk factors for cervical cancer -- Chapter 5. Secondary prevention with screening programs -- Chapter 6. Cancer prevention in developing countries -- Chapter 7. Cervical cytology and alternative methods of screening -- Chapter 8. Management of women with abnormal cytological results -- Chapter 9. Use of biomarkers in cervical cancer screening -- Chapter 10. Basic concepts of quality and accreditation in Health Care Services -- Chapter 11. Quality issues in screening test procedures -- Chapter 12. Quality issues in the cervical cancer screening program: indicators and standards -- Chapter 13. Instruction and training of personnel in a cervical cancer screening program -- Chapter 14. Universal hygienic measures and precautions for infection prevention in gynecological ambulatory centers and hospitals.
Subjects	Cervix uteri--Cancer--Diagnosis. Cervix uteri--Etiology. Cervix Uteri. Neoplasms--diagnosis. Neoplasms--etiology.
Notes	"Nova Biomedical." Includes bibliographical references and index.
Series	Cancer etiology, diagnosis and treatments Cancer etiology, diagnosis, and treatments.

A planet of viruses

LCCN	2010036742
Type of material	Book
Personal name	Zimmer, Carl, 1966-
Main title	A planet of viruses / Carl Zimmer.
Published/Created	Chicago; London: University of Chicago Press, c2011.
Description	x, 109 p.: col. ill.; 23 cm.
ISBN	9780226983356 (cloth: alk. paper) 0226983358 (cloth: alk. paper)
LC classification	QR360 .Z65 2011
Contents	"A contagious living fluid": tobacco mosaic virus -- Old companions -- The uncommon cold: rhinovirus -- Looking down from the stars: influenza virus -- Rabbits with horns: papillomavirus -- Everywhere, in all things -- The enemy of our enemy: bacteriophages -- Oceans of viruses: marine phages -- The infected genome: endogenous retroviruses -- The viral future -- The young scourge: human immunodeficiency virus -- Becoming an American: West Nile virus -- Predicting the next plague: severe acute respiratory syndrome -- The long goodbye: smallpox -- The alien in the watercooler: mimivirus.
Subjects	Viruses.
Notes	Includes bibliographical references (p. 97-101) and index.

A planet of viruses

LCCN	2015011313
Type of material	Book
Personal name	Zimmer, Carl, 1966- author.
Main title	A planet of viruses / Carl Zimmer.

Edition	Second edition.
Published/Produced	Chicago: The University of Chicago Press, [2015] ©2015
Description	x, 122 pages: illustrations (some color); 22 cm
ISBN	9780226294209 (paper: alk. paper) 022629420X (paper: alk. paper)
LC classification	QR360 .Z65 2015
Contents	Foreword / by Judy Diamond and Charles Wood -- Introduction -- "A contagious living fluid" -- Tobacco mosaic virus and the discovery of the virosphere -- Old companions -- The uncommon cold -- How rhinoviruses gently conquered the world -- Looking down from the stars -- Influenza's never-ending reinvention -- Rabbits with horns -- Human papillomavirus and infectious cancer -- Everywhere, in all things -- The enemy of our enemy -- Bacteriophages as viral medicine -- The infected ocean -- How marine phages rule the sea -- Our inner parasites -- Endogenous retroviruses and our virus-riddled genomes -- The viral future -- The young scourge -- Human immunodeficiency virus and the animal origins of diseases -- Becoming an American -- The globalization of West Nile virus -- Predicting the next plague -- Ebola virus and the many others like it -- The long goodbye -- The delayed oblivion of smallpox -- Epilogue.
Subjects	Viruses.
Notes	Includes bibliographical references (pages 111-116) and index.

Advances in cell and molecular diagnostics

LCCN	2017300284
Type of material	Book

Personal name	Raghavendra, Pongali, author.
Main title	Advances in cell and molecular diagnostics / Pongali Raghavendra, Thammineni Pullaiah.
Published/Produced	London, United Kingdom: Elsevier/Academic Press, [2018]
Description	xv, 281 pages: illustrations
ISBN	9780128136799 0128136790
Related names	Pullaiah, T., author.
Summary	Advances in Cell and Molecular Diagnostics brings the scientific advances in the translation and validation of cellular and molecular discoveries in medicine into the clinical diagnostic setting. It enumerates the description and application of technological advances in the field of cellular and molecular diagnostic medicine, providing an overview of specialized fields, such as biomarker, genetic marker, screening, DNA-profiling, NGS, cytogenetics, transcriptome, cancer biomarkers, prostate specific antigen, and biomarker toxicologies. In addition, it presents novel discoveries and clinical pathologic correlations, including studies in oncology, infectious diseases, inherited diseases, predisposition to disease, and the description or polymorphisms linked to disease states. This book is a valuable resource for oncologists, practitioners and several members of the biomedical field who are interested in understanding how to apply cutting-edge technologies into diagnostics and healthcare.
Contents	Chapter 1. Cellular and Molecular Diagnostics: An Introduction -- Chapter 2. RNA-Based Applications in Diagnostic and Therapeutics for

	Cancer -- Chapter 3. Advancements in Genetic Applications for Cellular and Molecular Diagnostics -- Chapter 4. Biomedical Imaging Role in Cellular and Molecular Diagnostics -- Chapter 5. Breast Cancer-Targeted Therapy Using Nanocarriers -- Chapter 6. Human Papillomavirus- and Epstein- Barr Virus-Caused Tumor Diagnosis and Therapy -- Chapter 7. Pathogen Identification Using Novel Sequencing Methods -- Chapter 8. Future of Cellular and Molecular Diagnostics: Bench to Bedside.
Subjects	Molecular diagnosis. Cells--Diagnostic use. Pathology, Molecular. Cytogenetic Analysis. Molecular Diagnostic Techniques. Molecular diagnosis. Pathology, Molecular.
Notes	Includes bibliographical references and index.

Adverse effects of vaccines: evidence and causality

LCCN	2012007052
Type of material	Book
Corporate name	Institute of Medicine (U.S.). Committee to Review Adverse Effects of Vaccines.
Main title	Adverse effects of vaccines: evidence and causality / Committee to Review Adverse Effects of Vaccines, Board on Population Health and Public Health Practice; Kathleen Stratton ... [et al.], editors.
Published/Created	Washington, D.C.: National Academies Press, c2012.
Description	xxvii, 865 p.: ill.; 24 cm.
ISBN	9780309214353 (hardcover)

	0309214351 (hardcover)
LC classification	RA638 .I5567 2012
Related names	Stratton, Kathleen R.
Contents	Approach -- Evaluating biological mechanisms of adverse events -- Measles, mumps, and rubella vaccine -- Varicella virus vaccine -- Influenza vaccine -- Hepatitis A vaccine -- Hepatitis B vaccine -- Human papillomavirus vaccine -- Diphtheria toxoid-, tetanus toxoid-, and acellular pertussis-containing vaccines -- Meningococcal vaccine -- Injection-related adverse events -- Concluding comments.
Subjects	Vaccination--Complications. Vaccines. Vaccines--adverse effects. Bacterial Infections--prevention & control. Causality. Virus Diseases--prevention & control.
Notes	Includes bibliographical references (p. 749-826) and index.

Adverse events following vaccination against human papillomavirus: results of the 2010 campaign in the Netherlands

LCCN	2012445309
Type of material	Book
Main title	Adverse events following vaccination against human papillomavirus: results of the 2010 campaign in the Netherlands / T.M. van 't Klooster ... et al.
Published/Created	Bilthoven: National Institute for Public Health and the Environment, Ministry of Health, Welfare and Sport, c2011.
Description	36 p.: ill.; 30 cm.
LC classification	QR189.5.P36 A38 2011

Related names	Klooster, T. M. van 't. Rijksinstituut voor Volksgezondheid en Milieu (Netherlands) Netherlands. Ministerie van Volksgezondheid, Welzijn en Sport.
Subjects	Papillomavirus Vaccines--Netherlands. Health Surveys--Netherlands. Papillomaviruses--Netherlands. Papillomavirus diseases--Netherlands.
Notes	Includes bibliographical references (p. 31-32). Abstracts in English and Dutch.
Series	RIVM letter report; 210012002/2011 RIVM letter report; 210012002/2011

AIDS sourcebook: basic consumer health information about the Human Immunodeficiency Virus (HIV) and Acquired Immunodeficiency Syndrome (AIDS), including facts about its origins, symptoms, stages, types, transmission, risk factors, and prevention, and featuring details about diagnostic testing, treatments, side effects, and co-occurring infections, such as Mycobacterium Avium Complex, Tuberculosis, Cryptococcal Meningitis, Toxoplasmosis, and Human Papillomavirus. Along with tips for Living with HIVAIDS, Information on Clinical Trials and Current Research, a Glossary of Related Terms, and a List of Resources for Additional Help and information.

LCCN	2015035127
Type of material	Book
Main title	AIDS sourcebook: basic consumer health information about the Human Immunodeficiency Virus (HIV) and Acquired Immunodeficiency Syndrome (AIDS), including facts about its origins, symptoms, stages, types, transmission, risk factors, and prevention, and featuring details about diagnostic testing, treatments, side effects,

	and co-occurring infections, such as Mycobacterium Avium Complex, Tuberculosis, Cryptococcal Meningitis, Toxoplasmosis, and Human Papillomavirus. Along with tips for Living with HIV/AIDS, Information on Clinical Trials and Current Research, a Glossary of Related Terms, and a List of Resources for Additional Help and information.
Edition	Sixth edition.
Published/Produced	Detroit, MI: Omnigraphics, [2016]
Description	xvii, 620 pages: illustrations; 24 cm.
ISBN	9780780813816 (hardcover: acid-free paper)
LC classification	RC606.64 .A337 2016
Related names	Jones, Keith.
Summary	"Provides basic consumer health information about transmission, testing, stages, and treatment of human immunodeficiency virus (HIV), with facts about prevention, related complications, and tips for living with HIV/AIDS. Includes index, glossary of related terms and directory of resources"-- Provided by publisher.
Subjects	AIDS (Disease)--Popular works.
Notes	Includes bibliographical references and index.
Series	Health reference series

Animal biotechnology: models in discovery and translation

LCCN	2013372403
Type of material	Book
Main title	Animal biotechnology: models in discovery and translation / edited by Ashish S. Verma, Amity Institute of Biotechnology, Amity University Uttar Pradesh, NOIDA (UP), India, Anchal Singh, Amity Institute of Biotechnology, Amity University Uttar Pradesh, NOIDA (UP), India.

Published/Produced	Amsterdam; Boston: Elsevier/AP, Academic Press is an imprint of Elsevier, 2014.
Description	xxix, 620 pages: illustrations; 28 cm
ISBN	9780124160026 (cloth) 0124160026 (cloth) 9780123914347 (ebk.) 0123914345 (ebk.)
LC classification	SF140.B54 A57 2014
Related names	Verma, Ashish S, editor of compilation. Singh, Anchal, editor of compilation.
Contents	Human diseases: in vivo and in vitro models. Drosophila: a model for biotechnologists / K. Ravi Ram and D. Kar Chowdhuri -- Animal models of tuberculosis / Devyani Dube, Madhu Gupta, Udita Agrawal and Suresh P. Vyas -- Animal models for neurodegenerative disorders / Hitomi Tsuiji and Koji Yamanaka -- Epigenetics and animal models: applications in cancer control and treatment / Mukesh Verma, Neelesh Agarwal and Mudit Verma -- Development of mouse models for cancer research / Amrita Datta adn Debasis Mondal -- Human papillomavirus (HPV): diagnosis and treatment / Mausumi Bharadwaj, Showket Hussain, Richa Tripathi, Neha Singh and Ravi Mehrotra -- Human DNA tumor viruses and oncogenesis / Pravinkumar Purushothaman and Subhash Chandra Verma -- Animal models for human disease / Mohammad Reza Khorramizadeh and Farshid Saadat -- HIV and antiretroviral drugs / Ashish Swarup Verma, Iqram Govind Singh, Ruby Bansal and Anchal Singh -- Animal biotechnology as a tool to understand and fight aging / Pawan Kumar Maurya -- Animal biotechnology: tools and

techniques. Multicellular spheroid: 3-D tissue culture model for cancer research / Suchit Khanna, Anant Narayan Bhatt and Bilikere S. Dwarakanath -- Animal tissue culture: principles and applications / Anju Verma -- Concepts of tissue engineering / Poonam Verma and Vipin Verma -- Nanotechnology and its applications to animal biotechnology / Ashok K. Adya and Elisabetta Canetta -- Antibodies: monoclonal and polyclonal / Anchal Singh, Sushmita Chaudhary, Ashima Agarwal and Ashish Swarup Verma -- Molecular markers: tool for genetic analysis / Avinash Marwal, Anurag Kumar Sahu and R.K. Gaur -- Gene expression: analysis and quantitation / Denys V. Volgin -- Ribotyping: a tool for molecular taxonomy / S.K. Kashyap, S. Maherchandani and Naveen Kumar -- Next generation sequencing and its applications / Anuj Kumar Gupta and U.D. Gupta -- Biomolecular display technology: a new tool for drug discovery / Madhu Biyani, Koichi Nishigaki, adn Manish Biyani -- In silico models: from simple networks to complex diseases / Debmalya Barh, Vijender Chaitankar, Eugenia Ch Yiannakopoulou, Emmanuel O Salawu, Sudhir Chowbinia, Preetam Ghosh and Vasco Azevedo -- Animal biotechnology: applications and concerns. Transgenic animals and their applications / Shet Masih, Pooja Jain, Rahsa El Baz and Zafar K. Khan -- Stem cells: a trek from laboratory to clinic to industry / Bhudev C. Das and Abhisek Tyagi -- Role of cytogenetics and molecular genetics in human health and medicine / Madhumita Roy Chowdhury and Sudhisha Dubey -- Antibodies

	and their applications / Fahim Halim Khan -- Vaccines: present status and applications / Dinesh K. Yadave, Neelam Yadav and Satyendra Mohan Paul Khurana -- Safety assessment of food derived from genetically modified crops / Premendra D. Dwivedi, Mukul Das, Sandeep Kumar and Alok Kumar Verma -- Nanotechnology and detection of microbial pathogens / Rishi Shanker, Gulshan Singh, Anurag Jyoti, Premendra Dhar Dwivedi and Surinder Pal Singh -- Biotechnological exploitation of marine animals / Surajit Das -- Herbal medicine and biotechnology for the benefit of human health / Priyanka Srivastava, Mithilesh Singh, Gautami Devi and Rakhi Chaturvedi -- Perspectives on the human genome / Aruna Kumar and Kailash C. Upadhyaya -- Ethical issues in animal biotechnology / Abhik Gupta.
Subjects	Animal biotechnology. Biomedical engineering. Genetic engineering. Human-animal relationships. Technology.
Notes	Includes bibliographical references and index.

Bladder cancer: etymology, diagnosis, and treatments

LCCN	2010015572
Type of material	Book
Main title	Bladder cancer: etymology, diagnosis, and treatments / William E. Nilsson, editor.
Published/Created	Hauppauge, N.Y.: Nova Science Publishers, 2010.
Description	xiii, 255 p.: ill. (some col.); 26 cm.

ISBN	9781617280023 (hardcover)
LC classification	RC280.B5 B638 2010
Related names	Nilsson, William E.
Contents	Recurrence, progression, adjuvant treatment and follow-up in non-muscle invasive bladder cancer / Bogdan Geavlete, Antoine G. van der Heijden, J. Alfred Witjes -- Conservative management of non-muscle invasive bladder cancer after bacillus calmette-guérin failure / Andrew J. Lightfoot, Henry M. Rosevear, Michael A. O'Donnell -- Recent advances in intravesical therapy for non-muscle invasive bladder cancer / A.H.Mostafid -- Genetic and molecular mechanisms in bladder cancer development / S. Gunes, R. Buyukalpelli, Z. Yegin -- Investigation of molecular proteins for urothelial carcinoma in the urinary bladder / Kazumasa Matsumoto ... [et al.] -- Robot-assisted laparoscopic radical cystectomy for bladder cancer / Ja Hyeon Ku, Hyeon Hoe Kim -- Therapeutic effect of immediate intravesical instillation of mitomycin C after transurethral resection in patients with low risk non-muscle-invasive bladder tumours / Jesus Fernandez-Gomez, Antonio Jalon-Monzon -- Detection and genotyping by microarray of human papillomavirus in tissue of bladder and bladder carcinoma biopsy / Antonio Sorlózano, Ana Isabel Moraga ... [et al.] -- Digital image analysis in research and diagnosis of urinary bladder cancer / A. Dulewicz, D. Pietka, P. Jaszczak -- Pleiotropic effects of a micronutrient mixture on critical parameters of bladder cancer / M.W. Roomi ... [et al.].
Subjects	Bladder--Cancer.

Bladder--Cancer--Diagnosis.
Bladder--Cancer--Treatment.
Urinary Bladder Neoplasms.

Notes Includes bibliographical references and index.

Bodies in transition in the health humanities: representations of corporeality

LCCN 2019010602
Type of material Book
Main title Bodies in transition in the health humanities: representations of corporeality / edited by Lisa M. DeTora and Stephanie M. Hilger.
Published/Produced London; New York: Routledge, 2019.
Description 1 online resource.
ISBN 9781351128742 (master)
LC classification RA776.5
Related names DeTora, Lisa, 1966- editor.
Hilger, Stephanie M. (Stephanie Mathilde), editor.
Contents Introduction: bodies and transitions in the health humanities / Lisa M. DeTora and Stephanie M. Hilger -- Enlightened wax works: viewing the anatomical woman in the Viennese Josephinum / Angelika Vybiral -- Epistemological anxiety: the case of Michel-Anne Drouart / Stephanie M. Hilger -- Charting intersex: intersex life-writing and the medical record / Katelyn Dykstra -- Narrating sex change in Iran: transsexuality and the politics of documentary film / Najmeh Moradiyan-Rizi -- Isolated bodies, isolated spaces: anorexia and bulimia in women's autobiographical narratives / Barbara Grüning -- Unseen enemies: neisseria, desire, and bodily discourse / Lisa M. DeTora -- The Human

	Papillomavirus vaccination: gendering the rhetorics of immunization in public health discourses / Jennifer A. Malkowski -- Bacteriology and modernity: phenomenology, bio-politics, ontology / Jens Lohfert Jørgensen -- Being-in-alien: the trinity of bodies in Prometheus (2012) and Alien: Covenant (2017) / Adnan Mahmutovic and Denise Ask Nunes -- Embodied transitions in Michel de Montaigne / Nora Martin Peterson and Peter Martin -- Witnessing illness: phenomenology of photographic self portraiture / Elizabeth Lanphier -- Disjunction and relationality in terminal illness writing / Yianna Liatsos -- Afterword: representation as a lens: teaching and researching in the health humanities / Carl Fisher.
Subjects	Human Body Attitude Disease Infection Gender Identity
Notes	Includes bibliographical references and index.
Additional formats	Print version: Bodies in transition in the health humanities London; New York: Routledge, 2019 9780815356066 (DLC) 2019009500

Bodies in transition in the health humanities: representations of corporeality

LCCN	2019009500
Type of material	Book
Main title	Bodies in transition in the health humanities: representations of corporeality / edited by Lisa M. DeTora and Stephanie M. Hilger.
Published/Produced	London; New York: Routledge, 2019.

ISBN	9780815356066 (hardback: alk. paper)
LC classification	RA776.5
Related names	DeTora, Lisa, 1966- editor. Hilger, Stephanie M. (Stephanie Mathilde), editor.
Contents	Introduction: bodies and transitions in the health humanities / Lisa M. DeTora and Stephanie M. Hilger -- Enlightened wax works: viewing the anatomical woman in the Viennese Josephinum / Angelika Vybiral -- Epistemological anxiety: the case of Michel-Anne Drouart / Stephanie M. Hilger -- Charting intersex: intersex life-writing and the medical record / Katelyn Dykstra -- Narrating sex change in Iran: transsexuality and the politics of documentary film / Najmeh Moradiyan-Rizi -- Isolated bodies, isolated spaces: anorexia and bulimia in women's autobiographical narratives / Barbara Grüning -- Unseen enemies: neisseria, desire, and bodily discourse / Lisa M. DeTora -- The Human Papillomavirus vaccination: gendering the rhetorics of immunization in public health discourses / Jennifer A. Malkowski -- Bacteriology and modernity: phenomenology, bio-politics, ontology / Jens Lohfert Jørgensen -- Being-in-alien: the trinity of bodies in Prometheus (2012) and Alien: Covenant (2017) / Adnan Mahmutovic and Denise Ask Nunes -- Embodied transitions in Michel de Montaigne / Nora Martin Peterson and Peter Martin -- Witnessing illness: phenomenology of photographic self portraiture / Elizabeth Lanphier -- Disjunction and relationality in terminal illness writing / Yianna Liatsos -- Afterword: representation as a lens:

	teaching and researching in the health humanities / Carl Fisher.
Subjects	Human Body Attitude Disease Infection Gender Identity
Notes	Includes bibliographical references and index.
Additional formats	Online version: Bodies in transition in the health humanities London; New York: Routledge, 2019 9781351128742 (DLC) 2019010602

Breaking the silence on cancer and sexuality: a handbook for healthcare providers

LCCN	2017060214
Type of material	Book
Personal name	Katz, Anne (Anne Jennifer), 1958- author.
Main title	Breaking the silence on cancer and sexuality: a handbook for healthcare providers / by Anne Katz.
Edition	Second edition.
Published/Produced	Pittsburgh, Pennsylvania: Oncology Nursing Society, [2018]
Description	1 online resource.
ISBN	9781635930177 ()
LC classification	RC262
Related names	Oncology Nursing Society, issuing body.
Contents	Introduction: why is sexuality important in the context of cancer? -- A primer in human sexual functioning -- Sexual health assessment -- Breast cancer -- Prostate cancer -- Gynecologic cancers -- Testicular cancer -- Colorectal cancer -- Human papillomavirus-related cancers -- Impact of other cancers on sexuality -- Adolescents and young

	adults with cancer -- Middle aged and older adults with cancer -- Patients with cancer at or near the end of life -- The effect of the cancer experience on couples -- Sexual minority patients -- Symptoms affecting sexuality -- Body image -- Sexuality counseling -- Research in sexuality -- Resources.
Subjects	Neoplasms--psychology Nursing Care--psychology Patients--psychology Sex Counseling Sexuality
Notes	Includes bibliographical references and index.
Additional formats	Print version: Katz, Anne (Anne Jennifer), 1958- author. Breaking the silence on cancer and sexuality Second edition. Pittsburgh, Pennsylvania: Oncology Nursing Society, [2018] 9781635930160 (DLC) 2017057621

Breaking the silence on cancer and sexuality: a handbook for healthcare providers

LCCN	2017057621
Type of material	Book
Personal name	Katz, Anne (Anne Jennifer), 1958- author.
Main title	Breaking the silence on cancer and sexuality: a handbook for healthcare providers / by Anne Katz, PhD, RN, FAAN.
Edition	Second edition.
Published/Produced	Pittsburgh, Pennsylvania: Oncology Nursing Society, [2018]
Description	ix, 293 pages: illustrations; 23 cm
ISBN	9781635930160 (pbk.) 1635930162 (pbk.)
LC classification	RC262 .K358 2018

Related names	Oncology Nursing Society, issuing body.
Contents	Introduction: why is sexuality important in the context of cancer? -- A primer in human sexual functioning -- Sexual health assessment -- Breast cancer -- Prostate cancer -- Gynecologic cancers -- Testicular cancer -- Colorectal cancer -- Human papillomavirus-related cancers -- Impact of other cancers on sexuality -- Adolescents and young adults with cancer -- Middle aged and older adults with cancer -- Patients with cancer at or near the end of life -- The effect of the cancer experience on couples -- Sexual minority patients -- Symptoms affecting sexuality -- Body image -- Sexuality counseling -- Research in sexuality -- Resources.
Subjects	Cancer--Psychosomatic aspects. Cancer--Patients--Sexual behavior. Cancer--Complications. Sex (Psychology) Neoplasms--psychology Nursing Care--psychology Patients--psychology Sex Counseling Sexuality
Notes	Includes bibliographical references and index.
Additional formats	Online version: Katz, Anne (Anne Jennifer), 1958- author. Breaking the silence on cancer and sexuality Second edition. Pittsburgh, Pennsylvania: Oncology Nursing Society, [2018] 9781635930177 (DLC) 2017060214

Canadian woman studies: an introductory reader

LCCN	2019410609
Type of material	Book

Main title	Canadian woman studies: an introductory reader / edited by Brenda Cranney and Sheila Molloy; preface by Andrea O'Reilly.
Edition	Third edition.
Published/Produced	Toronto, Canada: INANNA Publications and Education Inc., [2015] ©2015
Description	xvii, 616 pages; 23 cm
ISBN	9781771330602 (paperback) 1771330600
LC classification	HQ1453 .C366 2015
Related names	O'Reilly, Andrea, 1961- writer of preface. Cranney, Brenda, 1949- editor. Molloy, Sheila, 1950-, editor.
Summary	This reader brings together articles on themes and topics at the forefront of feminist inquiry and research previously published in one of Canada's oldest feminist journals, Canadian woman studies.
Contents	Section I. Feminist Perspectives. From Riot Grrrl to Radical: Reflections from a Working-Class Feminist / Gina Whitfield -- Feminism & Multiculturalism in Quebec: An-Other Perspective / Dolores Chew -- Gender-Based Analysis and Differing Worldviews / Cynthia D. Stirbys -- Teaching Sexual Assault: The Education of Canadian Law Students / Rosemary Carins Way and Daphne Gilbert -- Does Yes Mean Yes? Exploring Sexual Coercion in Normative Heterosexuality / Hedda Hakvåg -- Alternative Altars: Beyond Patriarchy and Priesthood and Towards Inclusive Spirituality, Governance and Activism Among Catholic Women Religious in Ontario / Christine Gervais -

- Reclaiming Our Spirituality: A Pedagogical Tool for Feminism and Activism / Njoki Nathani Wane.
Section II. Herstories. It's Time for Change! The World March of Women 2000 / Pam Kapoor -- Women's Words: Power, Identity and Indigenous Sovereignty / Patricia A. Monture -- Remaking Waves: The Quebec Women's Movement in the 1950s and 1960s / Cheryl Gosselin -- Troubling Herstory: Unsettling White Multiculturalism in Canadian Feminism / Mary-Jo Nadeau -- "We're Here, Standing at the Shoreline": Sylvia Hamilton's Intervention in the Nova Scotian Discourse of Belonging and Multicultural Citizenship / Sharon Morgan Beckford -- "Los Desaparecidos": The Madres of the Plaza de Mayo and the Reframing of the Victims / H.M. Fraser -- What Women Need Now from Police and Prosecutors: 35 Years of Working to Improve the Police Response for Women Escaping Male Violence / Louisa Russell -- Creating Trialogue: Women's Constitutional Activism in Canada / Marilou McPhedran -- The Canadian Disabled Women's Movement: From Where Have We Come / Pat Israel and Fra Odette.
Section III. Work-Economy. "With the Appropriate Qualifications": Aboriginal People and Employment Equity / Patti Doyle-Bedwell -- Organizing on the "Factory of Wheels": The Bus Riders' Union and Anti-Racist Feminism for the 21st Century / Fiona Jeffries -- On Being A Feminist Farmer / Jennifer deGroot -- Strangers in an Estranged World: Two Radical Feminists in the Academy / Geneviéve Pagé and Ève-Marie

Lampron -- Dueling for Dollars: Feminist Activism and Minimum Wage Coalition Politics / Joan Grace.
Section IV. Policy. Women and the Canadian Legal System: Examining Situations of Hyper-Responsibility / CAEFS-NWAC -- Welfare Policy: A Critical Site of Struggle for Women's Safety / Janet Mosher and Pat Evans -- Why Women Still Ain't Satisfied: Politics and Activism in Canadian Child Care / Martha Friendly -- Arbitration and Family Laws: Muslim Women Campaign to Eliminate the Use of Religious Laws in Legally-Binding Arbitration / Alia Hogben -- Consent and Coercion in the Law of Rape in South Africa: A Feminist Transformative Approach / Shereen W. Mills -- Response to Canada's Apology to Residential School Survivors / Beverley Jacobs.
Section V. Violence. Commemoration for the Montreal Massacre Victims / Ursula Franklin -- Breaking the Silence: Reclaiming Qur'anic Interpretations as a Tool for Empowerment and Liberatory Praxis for Dealing with Domestic Violence in Canadian Muslim Communities / Sabra Desai and Zehra Haffajee -- Moving Beyond Rape as a "Weapon of War": An Exploration of Militarized Masculinity and its Consequences / Caitlin Maxwell -- Stolen Sisters: A Human Rights Response to Discrimination and Violence Against Indigenous Women in Canada / Amnesty International -- Violence Against the Women of Juárez / María Guadalupe Morfín Otero -- Homophobic Sexist Violence in Canada:

Trends in the Experiences of Lesbian and Bisexual Women in Canada / Ellen Faulkner.
Section VI. Representation. Erasing Race: The Story of Reena Virk / Yasmin Jiwani -- Aiming for Better Than "Nobody Flinched": Notes on Oppression in Cancer Care / Christine Sinding, Lisa Barnoff, Patti McGillicuddy, Pam Grassau, and Fran Odette -- Simpering Outrage During an "Epidemic" of Fetal Alcohol Syndrome / Caroline L. Tait -- The Role of Montreal's Dykes on Mykes Radio Show / Marie-Clair MacPhee and Mél Hogan -- Nunavut: Whose Homeland, Whose Voices? / Isabel Altamirano-Jiménez -- N'tacinowin inna nah': Our Coming In Stories / Alex Wilson -- No Woman Left Covered: Unveiling and the Politics of Liberation in Multi-interculturalism / Tanisha Ramachandran.
Section VII. Health. Feminist Perspectives on Breast Cancer, Environment Health and Primary Prevention: The Case for the Precautionary Principle / Dorothy Goldin Rosenberg -- "It's Your Body But": Young Women's Narratives of Refusing Human Papillomavirus (HPV) Vaccination / Francesca Mancuso and Jessica Polzer -- Traditional Healing and Spirituality among Grenadian Women: A Source of Resistance and Empowerment / Patsy Sutherland -- Notokwe Opikiheet: "Old Lady Raised": Aboriginal Women's Reflections on Ethics and Methodologies in Health / Kim Anderson.
Section VIII. Activism and Resistance. Wa(i)ving Solidarity: Feminist Activists Confronting Backlash / Victoria Bromley and Aalya Ahmad -- Advocacy, Activism and Social Change for

	Women in Prison / Kim Pate -- The Pedagogies of Transnational Feminist Activism / Debbie Lunny -- The Feminist Pacifist and Antimilitant Movement in Colombia: The Experience of la Ruta Pacífica de las Mujeres / Diana Maria Montealegre M. -- ¡Escuche Las Krudas! Raw, Feminist Rap Music from Havana / Talia Wooldridge -- Local Activism, Global Feminisms and the Struggle Against Globalization / Angela Miles.
Subjects	Women--Canada. Feminism--Canada.
Notes	Includes articles previously published in the journal Canadian woman studies. Includes bibliographical references and index.
Additional formats	Issued also in electronic format. Canadian woman studies (2015) Third edition. Toronto, Canada: INANNA Publications and Education Inc., 2015. (CaOONL)20149084455

Cancer-free!: are you sure?

LCCN	2018043091
Type of material	Book
Personal name	Hrbacek, Jenny, 1961- author.
Main title	Cancer-free!: are you sure? / Jenny Hrbacek, RN.
Published/Produced	Lake Mary, Florida: Siloam, [2018]
Description	1 online resource.
ISBN	9781629995540 (ebook)
LC classification	RC270
Summary	"Roughly one out of every three people will be diagnosed with cancer in their lifetime. Breast cancer survivor nurse Jenny Hrbacek hopes to change that. This book will empower readers to

gain control of their health and stop cancer before it's unstoppable"-- Provided by publisher.

Contents My story -- What is cancer? -- Statistics: misconceptions of reality -- Conventional treatments and testing -- Chemosensitivity, genomic, and natural-agent sensitivity testing overview -- AMAS test -- Biocept: circulating tumor cell count and biomarker identification -- BreastSentryTM -- The Cancer Profile© -- Cologuard®: stool DNA home screening kit -- Colon health screening for occult blood in stool: home test kit -- Colonsentry® -- Earlycdt-lung test -- Human chorionic gonadotropin (HCG) test -- Ivygene® -- Nagalase test -- Oral cancer screening: oralid: -- Papanicolaou (PAP) and human papillomavirus (HPV) -- Prostate health index -- Research Genetic Cancer Center -- Thymidine kinase test -- Videssa Breast® test -- Biofocus® tests -- Cellsearch®: circulating tumor cell count test -- Caris molecular intelligence -- Foundationone®: comprehensive genomic profile -- Nagourney Cancer Institute: chemosensitivity testing -- Additional labs offering chemosensitivity testing -- Suggested testing uses -- Connecting with an integrative physician -- Functional testing overview -- Low-dose approach to chemotherapy -- Cancer: beat it, don't feed it -- The toolbox -- Closing.

Subjects Cancer--Early detection.

Cancer--Treatment.

Additional formats Print version: Hrbacek, Jenny, 1961- author. Cancer-free! Lake Mary, Florida: Siloam, [2018] 9781629995533 (DLC) 2018029556

Cancer-free!

LCCN	2018029556
Type of material	Book
Personal name	Hrbacek, Jenny, 1961- author.
Main title	Cancer-free! / Jenny Hrbacek, RN.
Published/Produced	Lake Mary, Florida: Siloam, [2018] ©2018
Description	xxviii, 303 pages: illustrations; 23 cm
ISBN	9781629995533 (trade paper) 1629995533 (trade paper) (e-book)
LC classification	RC270 .H73 2018
Summary	"Roughly one out of every three people will be diagnosed with cancer in their lifetime. Breast cancer survivor nurse Jenny Hrbacek hopes to change that. This book will empower readers to gain control of their health and stop cancer before it's unstoppable"-- Provided by publisher.
Contents	My story -- What is cancer? -- Statistics: misconceptions of reality -- Conventional treatments and testing -- Chemosensitivity, genomic, and natural-agent sensitivity testing overview -- AMAS test -- Biocept: circulating tumor cell count and biomarker identification -- BreastSentryTM -- The Cancer Profile© -- Cologuard®: stool DNA home screening kit -- Colon health screening for occult blood in stool: home test kit -- Colonsentry® -- Earlycdt-lung test -- Human chorionic gonadotropin (HCG) test -- Ivygene® -- Nagalase test -- Oral cancer screening: oralid: -- Papanicolaou (PAP) and human papillomavirus (HPV) -- Prostate health index -- Research Genetic Cancer Center -- Thymidine kinase test -- Videssa Breast® test --

	Biofocus® tests -- Cellsearch®: circulating tumor cell count test -- Caris molecular intelligence -- Foundationone®: comprehensive genomic profile -- Nagourney Cancer Institute: chemosensitivity testing -- Additional labs offering chemosensitivity testing -- Suggested testing uses -- Connecting with an integrative physician -- Functional testing overview -- Low-dose approach to chemotherapy -- Cancer: beat it, don't feed it -- The toolbox -- Closing.
Subjects	Cancer--Early detection. Cancer--Treatment.
Notes	Includes bibliographical references and index.
Additional formats	Online version: Hrbacek, Jenny, 1961- author. Cancer-free! Lake Mary, Florida: Siloam, [2018] 9781629995540 (DLC) 2018043091

Cancers in people with HIV and AIDS: progress and challenges

LCCN	2014942269
Type of material	Book
Main title	Cancers in people with HIV and AIDS: progress and challenges / Robert Yarchoan, editor.
Published/Produced	New York: Springer, [2014] ©2014
Description	xvi, 389 pages: illustrations (some color); 24 cm
ISBN	9781493908585 (alk. paper) 1493908588 (alk. paper)
LC classification	RC606.6 .C362 2014
Related names	Yarchoan, Robert, editor.
Contents	HIV-associated cancers -- Epidemiology of AIDS-defining malignancies -- Epidemiology of non-AIDS defining malignancies -- HIV cancers in resource-limited regions -- Kaposi's Sarcoma-associated Herpesvirus (KSHV) -- Epstein Barr

Virus (EBV) -- Human Papillomavirus (HPV) -- Merkel Cell Polymavirus (MCV) -- Presentation and pathogenesis of Kaposi's Sarcoma -- Management of Kaposi's Sarcoma -- Presentation and pathogenesis of HIV lymphomas -- Diffuse large B-cell lymphoma -- Burkitt and Burkitt-like lymphoma -- Primary effusion lymphoma -- AIDS-related central nervous system lymphoma -- Plasmablastic and other lymphomas -- Hodkin lymphoma -- Multicentric Castelman Disease -- Cervical cancer -- Anal cancer -- Other HPV-associated cancers -- Lung cancer in HIV infection -- Hepattocellular carcinoma in HIV-positive patients -- Merkel cell carcinoma and other HIV-associated skin cancers -- Conjuctival carcinoma -- Malignancies in children with HIV infection -- cART and supportive care -- Stem cell transplantation.

Subjects AIDS malignancies.
HIV infections--Complications.
Neoplasms.
AIDS-Related Opportunistic Infections.
Acquired Immunodeficiency Syndrome--complications.
HIV Infections--complications.
AIDS malignancies.
HIV infections--Complications.

Notes "This book is an expansion of the section, Opportunistic Malignancies, in the Encyclopedia of AIDS, editors-in-chief Thomas J. Hope, Douglas D. Richman, and Mario Stevenson."
Includes bibliographical references and index.

Case studies in infection control

LCCN	2017056613
Type of material	Book
Personal name	Chand, Meera, author.
Main title	Case studies in infection control / Meera Chand and John Holton.
Published/Produced	London; New York: Taylor & Francis Group, [2018]
ISBN	9780815345176
LC classification	RA643
Related names	Holton, John, MRCPath, author.
Contents	Anthrax -- Aspergillosis -- Burkholderia cepacia complex (BCC) in cystic fibrosis -- Campylobacter jejuni infection -- Clostridium difficile infection -- An outbreak of Cryptosporidia sp. associated with a public swimming pool -- Giardia outbreaks on ship -- HIV -- Human papillomavirus (HPV) -- A laboratory incident linked to exposure to botulinum toxin -- Legionella pneumophila infection -- Multidrug-resistant tuberculosis (MDR-TB) -- Measles: achieving national control of a vaccine preventable infection -- Four cases of MERS-COV -- A case of hospital-acquired MRSA -- Neisseria gonorrhoeae with high-level resistance to azithromycin -- Increased number of infections with plasmodium SPP during a period of sociopolitical instability -- A clinical incident linked to prion-associated disease -- An outbreak of Pseudomonas aeruginosa in a neonatal intensive care unit -- Rabies -- An outbreak of nontyphoid salmonellosis in the workplace -- A case of toxoplasmosis in pregnancy -- Viral haemorrhagic fever -- A case of vero cytotoxin-

	producing Escherichia coli (VTEC) -- A case of varicella-zoster virus in a maternity unit.
Subjects	Communicable Disease Control Infection Control
Form/Genre	Case Reports
Notes	Includes index.
Additional formats	Online version: Chand, Meera, author. Case studies in infection control London; New York: Taylor & Francis Group, [2018] 9780203733318 (DLC) 2017058395

Case studies in infection control

LCCN	2017058395
Type of material	Book
Personal name	Chand, Meera, author.
Main title	Case studies in infection control / Meera Chand and John Holton.
Published/Produced	London; New York: Taylor & Francis Group, [2018]
Description	1 online resource.
ISBN	9780203733318 ()
LC classification	RA643
Related names	Holton, John, MRCPath, author.
Contents	Anthrax -- Aspergillosis -- Burkholderia cepacia complex (BCC) in cystic fibrosis -- Campylobacter jejuni infection -- Clostridium difficile infection -- An outbreak of Cryptosporidia sp. associated with a public swimming pool -- Giardia outbreaks on ship -- HIV -- Human papillomavirus (HPV) -- A laboratory incident linked to exposure to botulinum toxin -- Legionella pneumophila infection -- Multidrug-resistant tuberculosis (MDR-TB) -- Measles: achieving national control

	of a vaccine preventable infection -- Four cases of MERS-COV -- A case of hospital-acquired MRSA -- Neisseria gonorrhoeae with high-level resistance to azithromycin -- Increased number of infections with plasmodium SPP during a period of sociopolitical instability -- A clinical incident linked to prion-associated disease -- An outbreak of Pseudomonas aeruginosa in a neonatal intensive care unit -- Rabies -- An outbreak of nontyphoid salmonellosis in the workplace -- A case of toxoplasmosis in pregnancy -- Viral haemorrhagic fever -- A case of vero cytotoxin-producing Escherichia coli (VTEC) -- A case of varicella-zoster virus in a maternity unit.
Subjects	Communicable Disease Control Infection Control
Form/Genre	Case Reports
Notes	Includes index.
Additional formats	Print version: Chand, Meera, author. Case studies in infection control London; New York: Taylor & Francis Group, [2018] 9780815345176 (DLC) 2017056613

Cell-free protein synthesis: methods and protocols

LCCN	2013957131
Type of material	Book
Main title	Cell-free protein synthesis: methods and protocols / edited by Kirill Alexandrov and Wayne A. Johnston.
Published/Created	New York: Humana Press: Springer, c2014.
Description	xi, 313 p.: ill.; 26 cm.
Links	Inhaltsverzeichnis. http://bvbr.bib-bvb.de:8991/F?func=service&doc_library=BVB01&local_base=BVB01&doc_number=027060738&line_num

ber=0001&func_code=DB_RECORDS&service_type=MEDIA
Publisher description http://www.loc.gov/catdir/enhancements/fy1410/2013957131-d.html
Table of contents only http://www.loc.gov/catdir/enhancements/fy1410/2013957131-t.html

ISBN	9781627037815 (alk. paper) 1627037810 (alk. paper)
LC classification	QP551 .C4472 2014 QH506 .M45 v.1118
Related names	Alexandrov, Kirill edt Johnston, Wayne A. edt
Contents	Production of eukaryotic cell-free lysate from Leishmania tarentolae / Wayne A. Johnston and Kirill Alexandrov -- Bioinformatics analysis and optimization of cell-free protein synt / Alexander A. Tokmakov ... [et al.] -- Cell-free expression screen to identify fusion tags for improved protein expression / Andrew Kralicek -- One-pot, microscale cell-free enzyme expression and screening / Aarthi Chandrasekaran and Anup K. Singh -- Cell-free translation of biofuel enzymes / Taichi E. Takasuka ... [et al.] -- Cloning-independent expression and screening of enzymes using cell-free protein synthesis systems / Yong-Chan Kwon, Jae-Kwang Song, and Dong-Myung Kim -- High-level cell-free production of membrane proteins with nanodiscs / Christian Roos ... [et al.] -- Cell-free protein-based enzyme discovery and protein-ligand interaction study / Sabrina Guillemer ... [et al.] -- Human cell extract-derived cell-free systems for virus synthesis / Tominari Kobayashi, Kodai Machida, and Hiroaki Imataka -- Cell-free protein synthesis

in microfluidic 96-well plates / Kirsten Jackson, Ruba Khnouf, and Z. Hugh Fan -- Preparation of multiple site-specific mutant proteins for NMR studies by PCR-Directed cell-free protein synthesis / Kiyoshi Ozawa and Ruhu Qi -- Site-specific incorporation of unnatural amino acids into proteins by cell-free protein synthesis / Kiyoshi Ozawa and Choy Theng Loh -- In vitro translation of papillomavirus authentic and codon-modified L1 capsid gene mRNAs in mouse keratinocyte cell-free lysate / Kong-Nan Zhao -- Optimized yeast cell-free lysate system for in vitro translation of human virus mRNA / Xiao Wang, Liang Zhao, and Kong-Nan Zhao -- In vitro translation-based protein kinase substrate identification / Szilvia K. Nagy and Tamás Mészáros -- Preparation of protein arrays using cell-free protein expression / Elizabeth A. Cook and Mingyue He -- Posttranscriptional control of protein synthesis in Drosophila S2 cell-free system / Motoaki Wakiyama and Shigeyuki Yokoyama -- Cell-free membrane protein expression / Tomomi Kimura-Soyema, Mikako Shirouzu, and Shigeyuki Yokoyama -- PURE System for Protein Production / Yoshihiro Shimizu ... [et al.] -- Cell-free protein synthesis system from insect cells / Toru Ezure, Takashi Suzuki, and Eiji Ando -- Cell-free expression platform for production of protein microarrays / Xristo Zárate and David W. Galbraith.

Subjects Proteins--Synthesis--Laboratory manuals.
Protein Biosynthesis.
Cell-Free System--physiology.
Methode.

	Proteinbiosynthese.
	Zellfreies System.
Form/Genre	Aufsatzsammlung.
Notes	Includes bibliographical references and index.
Series	Methods in molecular biology, 1064-3745; 1118
	Springer protocols, 1949-2448
	Methods in molecular biology (Clifton, N.J.); v. 1118.
	Springer protocols (Series) 1949-2448

Cervical cancer: methods and protocols

LCCN	2014952732
Type of material	Book
Main title	Cervical cancer: methods and protocols / edited by Daniel Keppler, Athena W. Lin.
Published/Created	New York: Humana Press, c2015.
Description	xiv, 413 p.: ill. (some col.); 27 cm.
ISBN	9781493920129 (hbk.: acid-free paper)
	149392012X (hbk.: acid-free paper)
LC classification	RC280.U8 C4722 2015
Related names	Keppler, Daniel, 1957- edt
	Lin, Athena W., 1962- edt
	Springer Science+Business Media, copyright holder.
Summary	Representing the most relevant procedures and technologies aiding the advance of the field of HPV-mediated carcinogenesis of the cervix and other anatomical regions of squamocolumnar transition, such as the anorectum, penis, and oropharynx, Cervical Cancer: Methods and Protocols compiles a detailed collection of practical chapters. The first half of the book covers HPV types, pathogenesis of cervical cancer (CxCA), prevention, and novel potential

drug targets, while the second half explores pathology, genomics, modeling of CxCA, and experimental therapeutic strategies. Written in the highly successful Methods in Molecular Biology series format, chapters include introductions to their respective topics, lists of the necessary materials and reagents, step-by-step, readily reproducible laboratory protocols, and tips on troubleshooting and avoiding known pitfalls. Authoritative and vital, Cervical Cancer: Methods and Protocols serves as a valuable resource to both bench scientists and clinicians who step into the realm of high-risk HPVs and CxCA for the first time or those who wish to learn novel approaches or expand their toolbox for the study of CxCA.

Contents Evolution and classification of oncogenic human papillomavirus types and variants associated with cervical cancer / Zigui Chen, Luciana Bueno de Freitas, and Robert D. Burk -- Real-time PCR approach based on SPF10 primers and the INNO-LiPA HPV genotyping extra assay for the detection and typing of human papillomavirus / M. Isabel Micalessi, Gaëlle A. Boulet, and Johannes Bogers -- Replication of human papillomavirus in culture / Eric J. Ryndock, Jennifer Biryukov, and Craig Meyers -- HPV binding assay to laminin-332/integrin [alpha]6[beta]4 on human keratinocytes / Sarah A. Brendle and Neil D. Christensen -- Methods to assess the nucleocytoplasmic shuttling of the HPV E1 helicase and its effects on cellular proliferation and induction of a DNA damage response / Michaël Lehoux, Amélie Fradet-

Turcotte, and Jacques Archambault -- Genetic methods for studying the role of viral oncogenes in the HPV life cycle / Jason M. Bodily -- Robust HPV-18 production in organotypic cultures of primary human keratinocytes / Hsu-Kun Wang, Thomas R. Broker, and Louise T. Chow -- High-throughput cellular assay to quantify the p53-degradation activity of E6 from different human papillomavirus types / David Gagnon and Jacques Archambault -- Retroviral expression of human cystatin genes in HeLa cells / Crystal M. Diep, Gagandeep Kaur, Daniel Keppler, and Athena W. Lin -- Molecular analysis of human papillomavirus virus-like particle activated langerhans cells in vitro / Andrew W. Woodham, Adam W. Woodham, Adam B. Raff, Diane M. Da Silva, and W. Martin Kast -- Selective silencing of gene target expression by siRNA expression plasmids in human cervical cancer cells / Oscar Peralta-Zaragoza [and eleven others] -- Silencing of E6/E7 expression in cervical cancer stem- like cells / Wenyi Gu, Nigel McMillan, and Chengzhong Yu -- Two-step procedure for evaluating experimentally induced DNA damage: texas red and comet assays / Gina J. Ferris, Lauren B. Shunkwiler, and Charles A. Kunos -- Measurement of deubiquitinating enzyme Activity via a suicidal HA-Ub-VS probe / Colleen Rivard and Martina Bazzaro -- Immunocytochemical analysis of the cervical pap smear / Terry K. Morgan and Michelle Berlin -- Diagnosis of HPV-negative, gastric-type adenocarcinoma of the endocervix / Edyta C. Pirog -- Targeting of the HPV-16 E7 protein by

RNA aptamers / Julia Dolores Toscano-Garibay, María Luisa Benítez-Hess, and Luis Marat Alvarez-Salas -- Use of MYBL2 as a novel candidate biomarker of cervical cancer / Cara M. Martin [and others] -- Fixation methods for the preservation of morphology, RNAs, and proteins in paraffin-Embedded human cervical cancer cell xenografts in mice / Yoko Matsuda and Toshiyuki Ishiwata -- Assessment of the HPV DNA methylation status in cervical lesions / Mina Kalantari and Hans-Ulrich Bernard -- MeDIP-on-chip for methylation profiling / Yaw-Wen Hsu, Rui-Lan Huang, and Hung-Cheng Lai -- Use of DBD-FISH for the study of cervical cancer progression / Elva I. Cortés-Gutiérrez [and four others] -- Quantitative and high-throughput assay of human papillomavirus DNA replication / David Gagnon, Amélie Fradet-Turcotte, and Jacques Archambault -- Native human papillomavirus production, quantification, and infectivity analysis / Jennifer Biryukov, Linda Cruz, Eric J. Ryndock, and Craig Meyers -- Functional analysis of HPV-like particle-activated langerhans cells in vitro / Lisa Yan, Andrew W. Woodham, Diane M. Da Silva, and W. Martin Kast -- Assessment of the radiation sensitivity of cervical cancer cell lines / Mayil Vahanan Bose and Thangarajan Rajkumar -- Mouse model of cervicovaginal papillomavirus infection / Nicolas Çuburu, Rebecca J. Cevio, Cynthia D. Thompson, and Patricia M. Day -- Establishment of orthotopic primary cervix cancer xenografts / Naz Chaudary, Karolina Jaluba, Melania Pintilie, and Richard P. Hill -- Generation

	of K14-E7/[delta]N87[beta] cat double transgenic mice as a model of cervical cancer / Gülay Bulut and Aykut Üren.
Subjects	Cervix uteri--Cancer. Uterine Cervical Neoplasms--Laboratory Manuals. Human Papillomavirus DNA Tests--Laboratory Manuals--methods. Cervix uteri--Cancer.
Form/Genre	Laboratory Manuals.
Notes	Includes bibliographical references and index.
Series	Methods in molecular biology, 1064-3745; 1249 Springer protocols, 1949-2448 Methods in molecular biology (Clifton, N.J.); v. 1249, 1064-3745 Springer protocols (Series), 1949-2448

Congenital and perinatal infections

LCCN	2017040392
Type of material	Book
Uniform title	Congenital and perinatal infections (Read)
Main title	Congenital and perinatal infections / edited by Jennifer S. Read, MD, MS, MPH, DTM & H, Clinical Professor of Epidemiology; Department of Epidemiology and Biostatistics; University of California at San Francisco, San Francisco, CA; Previously: Cheif, Epidemiology Section and Senior Medical, Epidemiologist, DB/DVBD/ NCEZID/OID, Centers for Disease Control and Prevention, San Juan, PR, Mark R. Schleiss, MD, Director, Division of Pediatric Infectious Diseases and Immunology; Codirector, Center for Infectious Diseases and Microbiology Translational Research; American Legion and

	Auxiliary Heart Foundation Research Endowed Professor; University of Minnesota, Minneapolis, MN.
Published/Produced	Oxford; New York: Oxford University Press, [2018]
Description	ix, 329 pages: illustrations; 25 cm
ISBN	9780190604813 (hardback: alk. paper)
LC classification	RC111 .C675 2018
Related names	Read, Jennifer S., editor.
	Schleiss, Mark R., editor.
Summary	"Infectious diseases that specifically target the developing fetus and newborn infant represent a special category with unique management challenges. This book provides state-of-the-art information on epidemiology, pathogenesis, diagnosis, treatment, and prevention of the most important viral, bacterial and parasitic pathogens that are transmitted from mother to infant"--Provided by publisher.
Contents	Cytomegalovirus -- Ebola and other filoviruses -- Enterovirus -- Hepatitis B -- Hepatitis C -- Hepatitis E -- Herpes simplex virus -- Human immunodeficiency virus type 1 -- Human papillomavirus -- HTLV -- LCMV -- Parvovirus -- Rubella -- Zika -- Chagas -- Toxoplasma gondii -- Chlamydia trachomatis -- Group B Streptococcus -- Listeria -- Syphilis -- Ureaplasma.
Subjects	Communicable Diseases--transmission
	Infectious Disease Transmission, Vertical
	Infant, Newborn, Diseases
	Fetal Diseases
	Pregnancy Complications, Infectious
Notes	Includes bibliographical references and index.

Diagnostic microbiology of the immunocompromised host

LCCN	2016020812
Type of material	Book
Main title	Diagnostic microbiology of the immunocompromised host / edited by Randall T. Hayden, Donna M. Wolk, Karen C. Carroll, Yi-Wei Tang.
Edition	2nd edition.
Published/Produced	Washington, DC: ASM Press, [2016] ©2016
Description	xx, 794 pages; 29 cm
ISBN	9781555819033 (hardcover)
LC classification	QR67 .D55 2016
Related names	Hayden, Randall T., editor. Wolk, Donna M., editor. Carroll, Karen C., editor. Tang, Yi-Wei, editor.
Contents	I. Overview of infections in the immunocompromised host -- Overview of infections in the immunocompromised host / Lesia K. Dropulic and Howard M. Lederman -- II. Laboratory diagnosis: approaches, interpretations, and limitations: infections by specific etiologic agents -- Human immunodeficiency virus / Wendy S. Armstrong, Jeannette Guarner, Colleen S. Kraft, and Angela M. Caliendo -- Chronic hepatitis B, C, and D in the immunocompromised patient / Bryan R. Cobb and Alexandra Valsamakis -- Cytomegalovirus / Veronica Dioverti and Raymund R. Razonable -- Epstein-Barr infection in the immunocompromised host / Andrew Nowalk and Michael Green -- Herpes simplex virus and Varicella-Zoster virus / Myron J. Levin, Adriana

Weinberg, and D. Scott Schmid -- Human herpesviruses 6A, 6B and 7 / Henri Agut, Pascale Bonnafous, and Agnes Gautheret-Dejean -- Human papillomavirus / Eileen M. Burd and Christina L. Dean -- Polyomaviruses / Linda Cook -- Adenovirus / Michael G. Ison and Randall T. Hayden -- Respiratory RNA viruses / Richard L. Hodinka -- Enteroviruses and parechoviruses / James J. Dunn -- Parvovirus B19 / Marie Louise Landry -- Filamentous fungi / Margaret V. Powers-Fletcher, Brian A. Kendall, Allen T. Griffin, and Kimberly E. Hanson -- Yeasts / Sean X. Zhang and Nathan P. Wiederhold -- Mycobacteria / Patricia J. Simner, Gail L. Woods, and Nancy L. Wengenack -- Aerobic actinomycetes of clinical significance / A. Brian Mochon, Den Sussland, and Michael A. Saubolle -- Parasites / Elitza S. Theel and Bobbi S. Pritt -- Aerobic bacteria and infections in the immunocompromised host / Geraldine Hall and Karen C. Carroll -- Selected topics in anaerobic bacteriology / Deirdre L. Church -- III. Laboratory diagnosis: approaches, interpretations and limitations: infections of specific organ systems -- Etiology, epidemiology and diagnosis of lower respiratory tract infections in immunocompromised patients / Karen C. Carroll and La'Tonzia L. Adams -- Genitourinary tract infections in immunocompromised patients / Odaliz Abreu Lanfranco and George J. Alangaden -- Gastrointestinal infections / Kevin Alby and Irving Nachamkin -- Central nervous system infections -- Andrea J. Zimmer, Victoria E. Burke, and Karen C. Bloch -- Bloodstream / Donna M.

	Wolk, Raquel M. Martinez, and Diana R. Hernandez -- Skin and soft tissue infection / Anne Spichler Moffarah, Mayar Al Mohajer, Bonnie L. Hurwitz, and David G. Armstrong -- IV. Special topics -- Prosthetic device infections / Raquel M. Martinez, Thomas R. Bowen, and Michael A. Foltzer -- Hospital-associated infections / Esther Babady -- Role of surgical pathologic diagnosis for immunocompromised hosts / Mary K. Klassen-Fischer and Ronald C. Neafie.
Subjects	Diagnostic microbiology--Technique. Immunological deficiency syndromes--Diagnosis. Opportunistic infections.
Notes	Includes index.
Additional formats	Online version: Diagnostic microbiology of the immunocompromised host 2nd edition. Washington, DC: ASM Press, [2016] 9781555819040 (DLC) 2016023282

Diagnostic virology protocols

LCCN	2010937421
Type of material	Book
Main title	Diagnostic virology protocols / edited by John R. Stephenson, Alan Warnes.
Edition	2nd ed.
Published/Created	New York, N.Y.: Humana Press, c2011.
Description	xiii, 470 p.: ill. (some col.); 27 cm.
Links	http://dx.doi.org/10.1007/978-1-60761-817-1 Connect to this title online. http://www.springer protocols.com/BookToc/doi/10.1007/978-1-60761-817-1
ISBN	9781607618164 (alk. paper) 1607618168 (alk. paper)

	9781607618171
	1607618176
LC classification	QR387 .D52 2011
Related names	Stephenson, John R.
	Warnes, Alan.
Contents	Detection of adenoviruses / Kanti Pabbaraju, Sallene Wong, and Julie D. Fox -- Alphaviruses / A.M. Powers and John T. Roehrig -- Detection of human caliciviruses in fecal samples by RT-PCR / Robert L. Atmar, Frederick H. Neill, and Françoise S. Le Guyader -- Detection of dengue virus / Nagesh K. Tripathi ... [et al.] -- Detection of enteroviruses from clinical specimens / Chit Laa Poh and Eng Lee Tan -- The detection of hepatitis viruses / Peter Karayiannis -- Molecular detection of herpesviruses / Harald H. Kessler and Sophie Heyszl -- HIV drug resistance testing / Patricia Cane -- Detection, quantification, and characterisation of HIV/SIV / Neil Berry, Carolina Herrera, and Martin Cranage -- Simultaneous molecular detection and confirmation of influenza AH5, with internal control / Joanna S. Ellis and Martin D. Curran -- Detection of measles, mumps, and rubella viruses / Graham Tipples and Joanne Hiebert -- Detection of high-risk mucosal human papillomavirus DNA in human specimens by a novel and sensitive multiplex PCR method combined with DNA microarray / Tarik Gheit and Massimo Tommasino -- The detection of parvoviruses / Sean Doyle -- Detection and characterization of polioviruses / Javier Martin -- Detection of human-pathogenic poxviruses / Andreas Kurth and Andreas Nitsche -- Lyssaviruses: special

emphasis on rabies virus and other members of the lyssavirus genus / Graeme Harkess and Anthony R. Fooks -- Simultaneous detection and differentiation of respiratory syncytial virus and other respiratory viral pathogens / Susan E. Sefers, Haijing Li, and Yi-Wei Tang -- Rotaviruses / Jim Gray and Miren Iturriza-Gómara -- Detection of Crimean-Congo hemorrhagic fever, hanta, and sandfly fever viruses by real-time RT-PCR / Sofi M. Ibrahim ... [et al.] -- Detection of SARS coronavirus / J.S. Malik Peiris and Leo L.M. Poon -- Detection of West Nile virus / Elizabeth B. Kauffman ... [et al.] -- Point of care testing: diagnosis outside the virology laboratory / Christopher C. Blyth, Robert Booy, and Dominic E. Dwyer -- Modelling emerging viral epidemics for public health protection / Steve Leach and ian Hall.

Subjects Diagnostic virology--Laboratory manuals.
Virus Diseases--diagnosis--Laboratory Manuals.
Viruses--isolation & purification--Laboratory Manuals.

Notes Includes bibliographical references and index.

Additional formats Also available on the World Wide Web.

Series Springer protocols
Methods in molecular biology, 1940-6029; 665
Springer protocols.
Methods in molecular biology (Clifton, N.J.); v. 665.

Dx/Rx. Human papilloma virus

LCCN 2010033531

Type of material Book

Personal name Dizon, Don S.

Main title	Dx/Rx. Human papilloma virus / Don S. Dizon, Ashley R. Stuckey, Michael L. Krychman.
Published/Created	Sudbury, MA: Jones & Bartlett Learning, c2012.
Description	vi, 70 p.: ill.; 21 cm.
ISBN	9780763781613 (pbk.) 0763781614 (pbk.)
LC classification	RC168.P15 D58 2012
Other title	Human papilloma virus
Related names	Stuckey, Ashley R. Krychman, Michael L.
Contents	Epidemiology and risk factors -- The HPV genome -- HPV and HIV -- HPV and diseases of the cervix -- HPV and lower genital disease -- HPV and head and neck cancer -- HPV and anal cancer -- HPV and penile cancer -- Preventing infection: the HPV vaccine.
Subjects	Papillomavirus diseases. Papillomavirus Infections. Alphapapillomavirus. Papillomavirus Vaccines.
Notes	Includes bibliographical references and index.

Epidemiologic studies in cancer prevention and screening

LCCN	2012951334
Type of material	Book
Main title	Epidemiologic studies in cancer prevention and screening / Anthony B. Miller, editor.
Published/Created	New York: Springer, 2013.
Description	xv, 418 p.: ill. (some col.), maps, charts; 24 cm.
ISBN	9781461455851 (alk. paper) 1461455855 (alk. paper)
LC classification	RA645.C3 E65 2013
Related names	Miller, A. B. (Anthony B.)

Summary	"Epidemiologic Studies in Cancer Prevention and Screening is the first comprehensive overview of the evidence base for both cancer prevention and screening. This book is directed to the many professionals in government, academia, public health and health care who need up-to-date information on the potential for reducing the impact of cancer, including physicians, nurses, epidemiologists, and research scientists. The main aim of the book is to provide a realistic appraisal of the evidence for both cancer prevention and cancer screening. In addition, the book provides an accounting of the extent programs based on available knowledge have impacted populations."--Back cover.
Contents	Prevention of Cancer -- Health Promotion Approaches to Reducing Cancer Incidence / Roy Cameron, Jon Kerner -- Preventing Cancer by Ending Tobacco Use / Neil Collishaw, Cynthia Callard -- Prevention of Occupationally Induced Cancer / Aaron Blair, Karin Hohenadel, Paul Demers, Loraine Marrett, Kurt Straif -- Human Papillomavirus Vaccination for the Prevention of Cervical and Other Related Cancers / F. Xavier Bosch -- Prevention of Cancers Due to Infection / Hideo Tanaka -- Applying Physical Activity in Cancer Prevention / Christine M. Friedenreich, Brigid M. Lynch, Annie Langley -- Cancer Prevention in the United States / Otis W. Brawley, Barnett S. Kramer -- The Role of Nutrition in Cancer Prevention / W. Philip T. James -- Chemoprevention of Cancer: From Nutritional Epidemiology to Clinical Trials / Mary Reid, James Marshall -- The Role of Hormonal Factors

in Cancer Prevention / David B. Thomas -- Controlling Environmental Causes of Cancer / Paolo Vineis -- A Historical Moment: Cancer Prevention on the Global Health Agenda / Andreas Ullrich -- Screening for Cancer -- Evidence-Based Cancer Screening / Anthony B. Miller -- Comprehensive Cervical Cancer Control: Strategies and Guidelines / Nathalie Broutet -- Screening for Cervical Cancer / Matti Hakama -- Screening for Colon Polyps and Cancer / Swati G. Patel, Dennis J. Ahnen -- Breast Cancer Screening / Anthony B. Miller -- Prostate Cancer Screening / Anthony B. Miller -- Applying Cancer Screening in the Context of a National Health Service / Julietta Patnick -- Impact of prevention and screening in populations -- Do International Trends in Cancer Incidence and Mortality Reflect Expectations from Cancer Screening? / Philippe Autier -- Using Mathematical Models to Inform Public Policy for Cancer Prevention and Screening / Natasha K. Stout, Michelle C. Dunn, J. Dik F. Habbema, David T. Levy, Eric J. Feuer -- Role of the Oncologist in Cancer Prevention / William Hryniuk -- Integrating Prevention and Screening for Lung Cancer into Clinical Practice / William Hocking -- Early Detection of Cancer in Asia (Including Australia) / Robert Burton, Cheng-Har Yip, Marilys Corbex -- Prevention and Screening for Cancer in Primary Health Care / Alan Katz, Jennifer Enns -- Finale: What Can We Expect from Cancer Prevention and Screening? / Anthony B. Miller.

Subjects Cancer--Epidemiology.

	Cancer--Prevention--Statistics.
	Cancer--Diagnosis--Statistics.
	Neoplasms--prevention & control.
	Mass Screening.
	Outcome Assessment (Health Care)
Notes	Includes bibliographical references and index.
Series	Statistics for biology and health, 1431-8776; 79
	Statistics for biology and health; 79. 1431-8776

Handbook on human papillomavirus: prevalence, detection and management

LCCN	2013935918
Type of material	Book
Main title	Handbook on human papillomavirus: prevalence, detection and management / Harris B. Smith, editor.
Published/Created	New York: Nova Biomedical, [2013]
Description	xviii, 548 pages: illustrations; 26 cm.
ISBN	9781626186729 (hbk.)
	1626186723 (hbk.)
LC classification	MLCM 2018/43085 (Q)
Related names	Smith, Harris B., editor.
Subjects	Papillomaviruses.
Notes	Includes bibliographical references and index.
Series	Virology research progress
	Public health in the 21st century

Head and neck cancer: a multidisciplinary approach

LCCN	2013012304
Type of material	Book
Uniform title	Head and neck cancer (2013)
Main title	Head and neck cancer: a multidisciplinary approach / editors, Louis B. Harrison, MD, FASTRO, Physician-in-Chief, Continuum Cancer

	Centers of New York, Chairman and Gerald J. Friedman Chair of Radiation Oncology, Beth Israel Medical Center and St. Luke's and Roosevelt Hospitals, Professor of Radiation Oncology and Otorhinolaryngology - Head and Neck Surgery, Albert Einstein School of Medicine, New York, New York, Roy B. Sessions, MD, FACS, Professor, Otolaryngology - Head and Neck Surgery, Medical University of South Carolina, Charleston, South Carolina, Adjunct Professor, Otolaryngology - Head and Neck Surgery, Georgetown University, Washington, District of Columbia, Merrill S. Kies. MD, Professor of Medicine, Thoracic/Head and Neck Medical Onncology, The University of Texas MD Anderson Cancer Center, Houston, Texas.
Edition	Fourth edition.
Published/Produced	Philadelphia, PA: Wolters Kluwer Health/Lippincott Williams & Wilkins, [2014] ©2014
Description	xxi, 908 pages: illustrations; 29 cm
ISBN	9781451144871 (hardcover: alk. paper) 1451144873 (hardcover: alk. paper)
LC classification	RC280.H4 H3845 2014
Related names	Harrison, Louis B., editor of compilation. Sessions, Roy B., editor of compilation. Kies, Merrill S., editor of compilation.
Contents	General principles of head and neck pathology -- Clinical evaluation of the head and neck cancer patient -- Prognostic factors in patients with head and neck cancer -- Radiologic evaluation of the head and neck cancer patient -- Novel radiation therapy techniques in the management of head

and neck cancer -- New approaches: robotics and endoscopic head and neck surgery -- Targeted agents in squamous cell carcinoma of the head and neck -- Multidisciplinary reconstruction of the head and neck: general principles -- Dental oncology and maxillofacial prosthetics -- Interdisciplinary symptom management -- Evaluation and rehabilitation of speech, voice, and swallowing functions after treatment of head and neck cancer -- Human papillomavirus--associated head and neck carcinoma -- Chemoprevention -- Management of the neck -- Metastatic cancer to the neck from an unknown primary site -- Cancer of the oral cavity -- Cancer of the oropharynx -- Cancer of the larynx -- Cancers of the hypopharynx and cervical esophagus -- Cancer of the nasal vestibule, nasal cavity paranasal sinuses, anterior skull base, and orbit: surgical management -- Cancer of the nasal cavity and paranasal sinus: radiation therapy and chemotherapy management -- Cancer of the nasopharynx -- Basal and squamous cell skin cancers -- Melanoma of the head and neck -- Management of salivary gland tumors -- Primary and recurrent disease of the lateral skull base -- Paragangliomas of the head and neck -- Cancer of the thyroid and parathyroid -- Soft tissue and bone sarcomas of the head and neck -- Systemic targeted therapy for recurrent and metastatic head and neck squamous cell cancer.

Subjects Neck--Cancer.
Head--Cancer.
Head and Neck Neoplasms--therapy.
Head and Neck Neoplasms--diagnosis.

Notes Includes bibliographical references and index.

Health disparities among under-served populations: implications for research, policy, and praxis

LCCN	2018418612
Type of material	Book
Main title	Health disparities among under-served populations: implications for research, policy, and praxis / edited by Sheri R. Notaro.
Edition	1st ed.
Published/Created	Bingley, UK: Emerald, ©2012.
Description	xxiii, 308 p.: ill.; 24 cm.
Links	1850-9999 http://www.emeraldinsight.com/
ISBN	9781781901021 1781901023
LC classification	RA448.4 .H423 2012
Related names	Notaro, Sheri R.
Summary	The health and well-being of our nation hangs in the balance. This anthology critically examines the devastating impact of race, class, and gender on the health and health care of African Americans, Latinos and American Indians, with particular focus on children and adolescents.
Contents	Putting tribal nations first: historical trends, current needs, and future directions in substance use prevention for American Indian and Alaska Native youths / Betty G. Brown, Julie A. Baldwin, Margaret L. Walsh -- Promotion of effective coping among adolescents of color / Jessie Kemmick Pintor, Carolyn Garcia, Ursula Reynoso -- Implementing trauma interventions in schools: addressing the immigrant and refugee experience / Olga Acosta Price [and others] -- Health disparities in African American maternal

and child health / Sheri R. Notaro -- Racial and global disparities in human papillomavirus infection and cervical cancer / Lauren D. Arnold, Vetta L. Sanders Thompson -- The Buffalo Lupus Project: a community-based participatory research investigation of toxic waste exposure and lupus / Edith M. Williams [and others] -- Teen pregnancy prevention among at-risk urban youth: improving parent-child connectedness / W. Douglas Evans [and others] -- Overrepresentation and disparate outcomes: examining health disparities based on sexual orientation in the context of homeless youths / Maurice N. Gattis -- Beyond individual-level sexual risk behaviors: HIV/AIDS and racial health disparities / Portia E. Adams -- Sociohistorical privilege in higher education: implications for African American student psychological adjustment, resiliency, and success at predominately white institutions / Matthew J. Taylor [and others] -- Can cultural competence have an impact on health disparities? / Brenda A. Battle -- Disparities in the prevalence of mental illness among black Americans / D. Lisa Cothran.

Subjects Minorities--Medical care--United States.
Minorities--Health and hygiene--United States.
Children of minorities--Medical care--United States.
Children of minorities--Health and hygiene--United States.
African Americans--Medical care.
Hispanic Americans--Medical care.
Indians of North America--Medical care.
Healthcare Disparities.

	Health Services Accessibility--organization & administration.
	Quality of Health Care--organization & administration.
	Minority Health.
	Ethnic Groups.
	Medically Underserved Area.
	Child.
	Adolescent.
	African Americans--Medical care.
	Children of minorities--Health and hygiene.
	Children of minorities--Medical care.
	Hispanic Americans--Medical care.
	Indians of North America--Medical care.
	Minorities--Health and hygiene.
	Minorities--Medical care.
	United States.
	United States.
Notes	Includes bibliographical references and index.
Series	Advances in education in diverse communities, 1479-358X; v. 9
	Advances in education in diverse communities; 9. 1479-358X

How pathogenic viruses think: making sense of virology

LCCN	2012003777
Type of material	Book
Personal name	Sompayrac, Lauren.
Main title	How pathogenic viruses think: making sense of virology / Lauren Sompayrac.
Edition	2nd ed.
Published/Created	Burlington, Mass.: Jones & Bartlett Learning, c2013.
Description	xiii, 167 pages: illustrations (some color); 28 cm

ISBN	9781449645793 1449645798
LC classification	QR482 .S65 2013
Related names	Sompayrac, Lauren. How pathogentic viruses work.
Contents	The organizing principle -- Host defenses -- The interferon defense system -- Influenza: a bait-and-switch virus -- Rhinovirus: a virus that surrenders -- Measles: a Trojan Horse virus -- Rotavirus: an undercover virus -- Adenovirus: a virus with a time limit -- Hepatitis A: a virus that detours -- Hepatitis B: a decoy virus -- Hepatitis C virus: an escape artist -- HTLV-I: a tribal virus -- HIV-1: an urban virus -- Herpes simplex: a virus that hides -- Human papillomavirus: a very quiet virus -- Emerging viruses -- Virus-associated cancer -- Vaccines -- Antiviral drugs.
Subjects	Viruses. Virus Diseases. Pathogenic viruses.
Notes	Rev. ed. of: How pathogenic viruses work / Lauren Sompayrac. c2002. Includes index.

HPV: diseases and disorders

LCCN	2014007191
Type of material	Book
Personal name	Parks, Peggy J., 1951-
Main title	HPV: diseases and disorders / Peggy J. Parks.
Published/Produced	San Diego, CA: ReferencePoint Press, Inc., [2015]
Description	96 pages: color illustrations, color map; 25 cm.
ISBN	9781601526908 (hardback: alk. paper) 1601526903 (hardback: alk. paper)

LC classification	RC168.P15 P367 2015
Variant title	Human papillomavirus Human papilloma virus
Subjects	Papillomavirus diseases--Juvenile literature.
Notes	Includes bibliographical references and index. Grades 9 to 12.
Series	Compact research series

HPV and cervical cancer: achievements in prevention and future prospects

LCCN	2012932967
Type of material	Book
Personal name	Borruto, Franco.
Main title	HPV and cervical cancer: achievements in prevention and future prospects / Franco Borruto, Marc de Ridder, editors.
Published/Produced	New York: Springer, [2012]
Description	xiv, 401 pages; 24 cm.
Links	Publisher description http://www.loc.gov/catdir/enhancements/fy1313/2012932967-d.html Table of contents only http://www.loc.gov/catdir/enhancements/fy1313/2012932967-t.html
ISBN	9781461419877
LC classification	RC168.P15 B67 2012
Variant title	Human papillomavirus and cervical cancer
Related names	De Ridder, Marc.
Subjects	Papillomavirus diseases. Cervix uteri--Cancer--Etiology. Papillomaviruses--Health aspects. Genital warts.
Notes	Includes bibliographical references and index.

HPV and other infectious agents in cancer: opportunities for prevention and public health

LCCN	2009022868
Type of material	Book
Main title	HPV and other infectious agents in cancer: opportunities for prevention and public health / Hans Krueger ... [et al.].
Published/Created	New York: Oxford University Press, c2010.
Description	xi, 547 p.: ill.; 24 cm.
ISBN	9780199732913 (alk. paper) 0199732914 (alk. paper)
LC classification	RC268.57 .H68 2010
Related names	Krueger, Hans.
Contents	Introduction: infection and cancer: an expanding paradigm -- Human papillomavirus: structure, transmission & occurrence -- Human papillomavirus: infection, natural history & carcinogenesis -- Human papillomavirus: associations with cervical cancer -- Human papillomavirus: associations with non-cervical cancer -- Human papillomavirus: detection of infection and disease -- Human papillomavirus: prevention of infection and disease -- Hepatitis viruses -- Helicobacter pylori -- Epstein-Barr virus -- Human herpesvirus type 8 -- Human T-cell lymphotropic virus type 1 -- Conclusion: infection and cancer: a paradigm shift.
Subjects	Viral carcinogenesis. Oncogenic viruses. Papillomaviruses. Papillomavirus diseases. Oncogenic Viruses. Alphapapillomavirus. Neoplasms--prevention & control.

	Neoplasms--virology.
	Tumor Virus Infections--prevention & control.
Notes	Includes bibliographical references and index.

Human papillomavirus (HPV) involvement in esophageal carcinogenesis

LCCN	2009033372
Type of material	Book
Personal name	Syrjänen, Kari J.
Main title	Human papillomavirus (HPV) involvement in esophageal carcinogenesis / Kari Syrjänen.
Published/Created	New York: Nova Biomedical Books, c2010.
Description	xiii, 96 p.; 23 cm.
ISBN	9781608762118 (pbk.)
LC classification	RC280.E8 S97 2010
Contents	Introduction -- Squamous cell papilloma -- Esophageal squamous cell carcinoma: epidemiology and risk factors -- Evidence of HPV involvement in esophageal cancer.
Subjects	Esophagus--Cancer--Etiology.
	Papillomaviruses.
	Papillomavirus diseases--Complications.
	Esophageal Neoplasms--etiology.
	Alphapapillomavirus.
	Papillomavirus Infections--complications.
Notes	Includes bibliographical references (p. 68-88) and index.
Series	Cancer etiology, diagnosis and treatments

Human papillomavirus (HPV)-associated oropharyngeal cancer.

LCCN	2015943436
Type of material	Book
Main title	Human papillomavirus (HPV)-associated oropharyngeal cancer.

Published/Produced	New York, NY: Springer Berlin Heidelberg, 2015.
Links	Publisher description http://www.loc.gov/catdir/enhancements/fy1509/2015943436-d.html Table of contents only http://www.loc.gov/catdir/enhancements/fy1509/2015943436-t.html
ISBN	9783319210995

Human papillomavirus: bench to bedside

LCCN	2014003258
Type of material	Book
Uniform title	Human papillomavirus (Ramírez-Fort)
Main title	Human papillomavirus: bench to bedside / volume editors, Marigdalia K. Ramírez-Fort, Houston, Tex., Farhan Khan, Houston, Tex., Peter L. Rady, Houston, Tex., Stephen K. Tyring, Houston, Tex.
Published/Produced	Basel; New York: Karger, [2014]
Description	vii, 271 pages: illustrations (chiefly color); 27 cm
ISBN	9783318025262 (hard cover: alk. paper)
LC classification	RC168.P15 H85 2014
Related names	Ramírez-Fort, Marigdalia K., editor of compilation. Khan, Farhan, editor of compilation. Rady, Peter L., editor of compilation. Tyring, Stephen K. (Stephen Keith), editor of compilation.
Contents	Human papillomavirus genomics: past, present and future / Harari, A., Chen, Z., Burk, R.D. -- The biology of human papillomaviruses / Nguyen, H.P., Ramirez-Fort, M.K., Rady, P.L. -- Viral oncogenesis / Doan, H.Q., Ramirez-Fort, M., Rady, P. -- Pathogenesis of infection by human papillomavirus / Brendle, S.A., Bywaters, S.M.,

Christensen, N.D. -- Host responses to infection with human papillomavirus / Stanley, M.A.; Sterling, J.C. -- Cutaneous human papillomavirus infection: manifestations and diagnosis / Tschandl, P., Rosendahl, C., Kittler, H. -- Epidermodysplasia verruciformis / Burger, B., Itin, P.H. -- Human papillomavirus infections of the oral mucosa and upper respiratory tract / Nguyen, H.P., McNiece, K.L., Duong, A.A., Khan, F. -- Human papillomavirus and immunosuppression / Wieland, U., Kreuter, A., Pfister, H. -- Laboratory diagnosis of human papillomavirus infection / Ikenberg, H. -- Management of cutaneous human papillomavirus infection: pharmacotherapies / Ramirez-Fort, M.K.; Au, S.-C., Javed, S.A., Loo, D.S. -- Management of cutaneous human papillomavirus infection: surgery / Ramirez-Fort, M.K., Sam, H., Manders, E.K. -- Management of cutaneous human papillomavirus infection: management of cutaneous human papillomavirus infection in immunocompromised patients / Varada, S., Posnick, M., Alessa, D., Ramirez-Fort, M.K. -- Anal and colon cancer / Yen Moore, A., Tong, L.X., Moore, T. -- Management of human papillomavirus-related head and neck cancer / Coughlin, A.M., Qiu, S., Underbrink, M.P. -- Vaccines and immunization against human papillomavirus / Christensen, N.D., Budgeon, L.R.

Subjects Papillomavirus Infections.
Papillomaviridae.

Notes Includes bibliographical references and index.

Series	Current problems in dermatology, 1421-5721; vol. 45 Current problems in dermatology, 1421-5721; v. 45.

Human papillomavirus infections: from laboratory to clinical practice

LCCN	2012939577
Type of material	Book
Personal name	Cobo, Fernando.
Main title	Human papillomavirus infections: from laboratory to clinical practice / Fernando Cobo.
Published/Created	Cambridge: Woodhead Pub., 2012.
Description	xviii, 149 p.: ill. (some col.); 24 cm.
ISBN	9781907568749 (hbk.) 1907568743 (hbk.) 9781908818171 (ebk.) 1908818174 (ebk.)
LC classification	QR406 .C63 2012
Summary	Human papillomavirus (HPV) is one of the most common causes of sexually transmitted diseases worldwide, both in men and women. This book provides the scientific background needed to understand the natural history and pathogenesis of HPV infection and offers discussion of its clinical features, diagnosis, treatment and prevention. It begins with chapters covering the epidemiology, virology, history and transmission of the infection, as well as its pathogenesis and clinical features. Following a discussion of the relationship between HPV and cancer, chapters in the second half of the book look at diagnosis, testing and treatment. The book concludes with

	detailed coverage of the prevention of HPV through worldwide vaccination programmes.
Contents	Introduction to epidemiological data -- Basic virology: structure and nomenclature -- Natural history and transmission: anatomic distribution - Pathogenesis and clinical features -- HPV infection and association with cancer -- Diagnosis of HPV infection -- Clinical utility of HPV testing -- Treatment -- Prevention: HPV vaccines.
Subjects	Papillomaviruses.
Notes	Includes bibliographical references and index.
Series	Woodhead Publishing series in biomedicine; no. 14

Miller's review of critical vaccine studies: 400 important scientific papers summarized for parents and researchers

LCCN	2015025036
Type of material	Book
Personal name	Miller, Neil Z., author.
Main title	Miller's review of critical vaccine studies: 400 important scientific papers summarized for parents and researchers / Neil Z. Miller.
Published/Produced	Santa Fe, New Mexico: New Atlantean Press, [2016]
Description	334 pages; 23 cm
ISBN	9781881217404 (pbk)
LC classification	RA638 .M54 2016
Variant title	Review of critical vaccine studies
Contents	Vaccination schedules -- Thimerosal (mercury) -- Aluminum -- Influenza -- Pertussis mutations -- Pathogen evolution and imperfect vaccines -- Strain replacement, haemophilus influenzae -- Strain replacement, pneumococcal disease -- Human papillomavirus (HPV) -- Measles and

	MMR -- Chickenpox and shingles -- Polio, hepatitis B, and rotavirus -- Allergies -- Seizures -- Diabetes -- Bleeding disorders -- Premature and low birth weight infants -- Hexavalent vaccines and sudden infant death (SIDS) -- Cancer and natural infections - Vitamin A and measles -- Vitamin D and influenza -- Non-vaccination by doctors and nurses -- Education level of non-vaccinating parents -- Conflicts of interest, false studies, and industry control.
Subjects	Vaccination. Vaccines. Biomedical Research. Child. Safety.
Notes	Includes bibliographical references and index.

Nose and viral cancer: etiology, pathogenesis and treatment

LCCN	2009035168
Type of material	Book
Main title	Nose and viral cancer: etiology, pathogenesis and treatment / Aloisio Medeiros and Carlitos Veloso, editors.
Published/Created	New York: Nova Science Publishers, c2010.
Description	xvi, 462 p.: ill. (some col.); 27 cm.
ISBN	9781607417354 (hardcover: alk. paper) 1607417359 (hardcover: alk. paper)
LC classification	RC268.57 .N67 2010
Related names	Medeiros, Aloisio. Veloso, Carlitos.
Contents	Human papillomavirus (HPV) involvement in esophageal carcinogenesis / Kari Syrjänen -- Viral cancers: contribution of cytologic tools to the diagnosis and management / Dilip K. Das --

Gammaherpesviruses and oncogenesis / M. Kúdelová and J. Rajčáni -- Sinonasal intestinal type adenocarcinoma: pathogenesis, diagnosis and treatment / Annarita Palomba ... [et al.] -- Pulmonary and nasal adenocarcinomas induced by ovine retroviruses / Naoyoshi Maeda and Hung Fan -- A voucher scheme approach to screening for cervical cancer: the Nicaraguan experience / Micol Salvetto and Vivian Alvarado -- Human papilloma virus: etiology, pathogenesis and prevention / S. Juan Carlos Roa ... [et al.] -- The forehead donor side of flaps in nasal reconstruction / Vasilios K. Thomaidis -- The oral cavity: a review of common lesions / Peter F. Koltz, Charles Butler and Samuel J. Lin -- Viral hepatitis and liver cancer / Yong-Song Guan, Qing He and Qing Zou -- Osteomas of the paranasal sinuses / Olaf Zagólski -- Rare nasal malignancies: lymphoepithiliomas, primary non-hodgkin lymphomas and mixed tumours, diagnostic and treatment challenges / Jiannis Hajiioannou ... [et al.] -- Occupational wood dust exposure and nasal sinus cancers: a review / Luc Fontana -- Reconstruction of nasal defects following nose cancer ablation / Raúl González-García ... [et al.] -- Nasopharyngeal carcinoma / Mazita Ami ... [et al.] -- Nasopharyngeal carcinoma, an Epstein-Barr-virus-related cancer, a summary of reports from Thailand / Viroj Wiwanitkit -- Chloroacetanilide-induced nasal carcinogenesis in rats / Mary Beth Genter and Charles Breckenridge.

Subjects Viral carcinogenesis.
Nose--Cancer.

	Nose Neoplasms--etiology.
	DNA Tumor Viruses.
	Nose Neoplasms--therapy.
	Tumor Virus Infections--etiology.
	Tumor Virus Infections--therapy.
Notes	Includes bibliographical references and index.
Series	Cancer etiology, diagnosis and treatments series Cancer etiology, diagnosis, and treatments.

Not quite a cancer vaccine: selling HPV and cervical cancer

LCCN	2017015420
Type of material	Book
Personal name	Gottlieb, S. D. (Samantha D.) author.
Main title	Not quite a cancer vaccine: selling HPV and cervical cancer / S.D. Gottlieb.
Published/Produced	New Brunswick: Rutgers University Press, [2018]
Description	vii, 200 pages: illustration; 24 cm
ISBN	9780813587783 (cloth) 9780813587776 (paperback)
LC classification	QR189.5.P36 G69 2018
Summary	"In Not Quite a Cancer Vaccine, medical anthropologist S.D. Gottlieb explores how the vaccine Gardasil--developed against the most common sexually-transmitted infection, human papillomavirus (HPV)--was marketed primarily as a cervical cancer vaccine. Gardasil quickly became implicated in two pre-existing debates--about adolescent sexuality and pediatric vaccinations more generally. Prior to its market debut, Gardasil seemed to offer female empowerment, touting protection against HPV and its potential for cervical cancer. Gottlieb questions the marketing pitch's vaunted promise and asks why vaccine marketing unnecessarily

	gendered the vaccine's utility, undermining Gardasil's benefit for men and women alike. This book demonstrates why in the ten years since Gardasil's U.S. launch its low rates of public acceptance have their origins in the early days of the vaccine dissemination. Not Quite a Cancer Vaccine addresses the on-going expansion in U.S. healthcare of patients-as-consumers and the ubiquitous, and sometimes insidious, health marketing of large pharma"-- Provided by publisher.
Subjects	Papillomavirus vaccines. Cervix uteri--Cancer--Prevention. Medical / Public Health. Social Science / Disease & Health Issues. Health & Fitness / Women's Health. Medical / Immunology. Health & Fitness / Diseases / Cancer.
Notes	Includes bibliographical references and index.
Additional formats	Online version: Gottlieb, S. D. (Samantha D.) author. Not quite a cancer vaccine New Brunswick: Rutgers University Press, [2018] 9780813587790 (DLC) 2017021033

Pharmacoeconomics: from theory to practice

LCCN	2009032763
Type of material	Book
Main title	Pharmacoeconomics: from theory to practice / editor, Renee J.G. Arnold.
Published/Created	Boca Raton, FL: Taylor & Francis, c2010.
Description	xiv, 243 p.: ill.; 25 cm.
ISBN	9781420084221 (hardcover: alk. paper)
LC classification	RS100 .P433 2010
Related names	Arnold, Renee J. G.

Contents Introduction to pharmacoeconomics / William F. McGhan -- Decision modeling techniques / Mark S. Roberts and Ken Smith -- Cost of illness / Renée J.G. Arnold -- Markov modeling in decision analysis / J. Robert Beck -- Retrospective database analysis / Renée J.G. Arnold and Sanjeev Balu -- What is cost-minimization analysis? / Alan Haycox -- Cost-effectiveness analysis / Kenneth J. Smith and Mark S. Roberts -- Budget impact analysis / Lieven Annemans -- Cost-utility analysis: a case study of a quadrivalent human Papillomavirus vaccine / Erik J. Dasbach, Ralph P. Insinga, and Elamin H. Elbasha -- Some problems/assumptions in pharmacoeconomic analysis / Stuart Birks -- Patient-reported outcome measures / Dianne Bryant, Gordon Guyatt, and Renée J.G. Arnold -- Sensitivity analysis / Maarten J. Postma -- Use of pharmacoeconomics in drug reimbursement in Australia, Canada, and the United Kingdom: what can we learn from international experience? / Michael Drummond and Corinna Sorenson -- Pharmacoeconomics in disease management: practical applications and persistent challenges / Ryung Suh and David Atkins -- Computer aided decision making from drug discovery to pharmacoeconomics / Sean Ekins and Renée J.G. Arnold -- Speculations on the future challenges and value of pharmacoeconomics / J. Jaime Caro, Denis Getsios, and Rachael L. Fleurence.

Subjects Drugs--Cost effectiveness.
Pharmacy--Economic aspects.
Decision making.
Economics, Pharmaceutical.

	Costs and Cost Analysis.
	Decision Making.
	Outcome Assessment (Health Care)--economics.
Notes	Includes bibliographical references and index.
Series	Drug discovery series; 13
	Drug discovery series; 13.

Plant-derived pharmaceuticals: principles and applications for developing countries

LCCN	2014006676
Type of material	Book
Main title	Plant-derived pharmaceuticals: principles and applications for developing countries / edited by Kathleen L. Hefferon, Cornell University, Ithaca, New York, USA
Published/Produced	Boston, MA: CABI, 2014.
Description	x, 170 pages: illustrations, maps; 24 cm.
ISBN	9781780643434 (alk. paper)
LC classification	RS164 .P7273 2014
Related names	Hefferon, Kathleen L.
Contents	Introduction, and the promise of a plant-derived vaccine for HBV / Kathleen Hefferon -- Protein body-inducing fusions for recombinant protein production in plants / Sonia Gutiérrez and Rima Menassa -- Plant cell culture expression systems / Somen Nandi and Karen McDonald -- Virus expression systems / Qiang Chen and Huafang Lai -- Plant virus-like particles / Ann E. Meyers, Inga I. Hitzeroth, and Edward P. Rybicki -- Expression of the capsid protein of human papillomavirus in plants as an alternative for the production of vaccines / José Francisco Castillo Esparza, Alberto Monroy García and Miguel Angel Gómez Li -- Patenting of plant-made recombinant

pharmaceuticals and access in the developing world / Pascal M.W. Drake, Sonia Sadone, and Harry Thangaraj -- Case study 1. Rabies / Elizabeth Loza-Rubio, Edith Rojas-Anaya -- Case study 2. Plant-made HIV vaccines and neutralizing antibodies / Carla Marusic and Marcello Donini -- Case study 3. The search for a plant-made vaccine for pandemic influenza virus / Kathleen Hefferon.

Subjects	Materia medica, Vegetable--Developing countries. Medicinal plants--Biotechnology--Developing countries. Pharmaceutical biotechnology--Developing countries. Vaccines--Biotechnology--Developing countries. Monoclonal antibodies--Biotechnology--Developing countries.
Notes	Includes bibliographical references and index.
Series	CABI biotechnology series; 2 CABI biotechnology series; 2.

President's Cancer Panel Meeting: transcript of proceedings, April 23-24, 2013, Hilton Miami Downtown, Miami, FL.

LCCN	2014406406
Type of material	Book
Corporate name	United States. President's Cancer Panel. Meeting (2013: Miami, Fla.)
Main title	President's Cancer Panel Meeting: transcript of proceedings, April 23-24, 2013, Hilton Miami Downtown, Miami, FL.
Published/Produced	[Rockville, Md.?]: Department of Health & Human Services, NIH, National Cancer Institute, [2013]

Description	162 pages; 28 cm
LC classification	RC168.P15 U57 2013
Variant title	Subtitle on cover: Global HPV vaccination: opportunities and challenges
Subjects	Papillomavirus diseases – Vaccination -- Congresses.

Prophylaxis and early detection of HPV-related neoplasia

LCCN	2012007708
Type of material	Book
Main title	Prophylaxis and early detection of HPV-related neoplasia / volume editor, Herbert Pfister.
Published/Created	Basel; New York: Karger, c2012.
Description	viii, 139 p.: col. ill.; 26 cm.
ISBN	9783805599641 (alk. paper) 9783805599658 (e-ISBN)
LC classification	RC168.P15 P76 2012
Related names	Pfister, Herbert, 1950-
Contents	Virology and pathogenesis / Pfister, H. -- Epidemiology of mucosotropic papillomaviruses / Gissmann, L. -- Condylomata acuminata / Gross, G. -- Recurrent respiratory papillomatosis / Wittekindt, C.; Wagner, S.; Klussmann, J.P. -- Cervical cancer and intraepithelial neoplasia / Schneider, A. ... [et al.] -- Cancer and intraepithelial neoplasia of the vulva and vagina / Petry, K.U. -- Penile cancer and penile intraepithelial neoplasia / Gross, G. -- Anal cancer and intraepithelial neoplasia / Wieland, U.; Kreuter, A. -- Tonsillar cancer / Klussmann, J.P.; Wagner, S.; Wittekindt, C. -- Cytology / Ikenberg, H. -- Concepts of biomarker development / von Knebel Doeberitz, M.; Vinokurova, S.; Reuschenbach, M. -- Detection of human

	papillomavirus DNA and RNA / Ikenberg, H. -- Human papillomavirus testing in primary screening for cervical cancer / Petry, K.U. -- Primary prevention by vaccination / Kaufmann, A.M.; Gissmann, L.
Subjects	Papillomavirus Infections. Early Detection of Cancer. Neoplasms--immunology. Neoplasms--virology. Papillomaviridae.
Notes	Includes bibliographical references and indexes.
Series	Monographs in virology, 0077-0965; v. 28 Monographs in virology, 0077-0965; v. 28.

Questions & answers about human papilloma virus (HPV)

LCCN	2009052706
Type of material	Book
Personal name	Dizon, Don S.
Main title	Questions & answers about human papilloma virus (HPV) / Don S. Dizon, Michael L. Krychman.
Published/Created	Sudbury, Mass.: Jones and Bartlett Publishers, c2011.
Description	x, 101 p.; 23 cm.
ISBN	9780763781620 (pbk.: alk. paper) 0763781622 (pbk.: alk. paper)
LC classification	RC168.P15 D59 2011
Variant title	Questions and answers about human papilloma virus (HPV)
Related names	Krychman, Michael L.
Subjects	Papillomavirus diseases--Miscellanea. Papillomaviruses--Miscellanea. Generative organs--Infections--Miscellanea.
Notes	Includes index.

Series	100 questions & answers 100 questions & answers.

Sexual health information for teens: health tips about sexual development, reproduction, contraception, and sexually transmitted infections: including facts about puberty, sexuality, birth control, hiv/aids, human papillomavirus, chlamydia, gonorrhea, herpes, and more

LCCN	2011007437
Type of material	Book
Main title	Sexual health information for teens: health tips about sexual development, reproduction, contraception, and sexually transmitted infections: including facts about puberty, sexuality, birth control, hiv/aids, human papillomavirus, chlamydia, gonorrhea, herpes, and more / edited by Elizabeth Magill.
Edition	3rd ed.
Published/Created	Detroit, Mich.: Omnigraphics, c2011.
Description	xiv, 428 p.: ill.,; 24 cm.
ISBN	9780780811553 (hardcover: alk. paper)
LC classification	RA777 .S47 2011
Related names	Magill, Elizabeth, 1971-
Summary	"Provides basic consumer health information for teens about puberty, sexuality, reproductive health, contraception, and prevention of sexually transmitted diseases. Includes index, resource information, and recommendation for further reading"--Provided by publisher.
Subjects	Teenagers--Health and hygiene. Sexual health. Reproductive health. Puberty. Sexually transmitted diseases--Prevention.

	Sex instruction for teenagers.
Notes	Includes bibliographical references (p. [399]-410) and index.
Series	Teen health series

Sexual health information for teens: health tips about sexual development, reproduction, contraception, and sexually transmitted infections including facts about puberty, sexuality, birth control, HIVAIDS, human papillomavirus, chlamydia, gonorrhea, herpes, and more.

LCCN	2015042291
Type of material	Book
Main title	Sexual health information for teens: health tips about sexual development, reproduction, contraception, and sexually transmitted infections including facts about puberty, sexuality, birth control, HIV/AIDS, human papillomavirus, chlamydia, gonorrhea, herpes, and more.
Edition	Fourth edition.
Published/Produced	Detroit, MI: Omnigraphics, [2016]
Description	xii, 355 pages: illustrations; 25 cm.
ISBN	9780780813854 (hardcover: acid-free paper)
LC classification	RA777 .S47 2016
Related names	Omnigraphics, Inc.
Summary	"Provides basic consumer health information for teens about puberty, development, and sexuality, maintaining sexual health, and preventing pregnancy and sexually transmitted diseases. Includes index and resource information"--Provided by publisher.
Subjects	Teenagers--Health and hygiene.
	Sexual health.
	Reproductive health.
	Puberty.

	Sexually transmitted diseases--Prevention.
	Sex instruction for teenagers.
Notes	Includes bibliographical references and index.
	Grade 9 to 12.
Additional formats	Online version: Sexual health information for teens Fourth edition. Detroit, MI: Omnigraphics, Inc., [2016] 9780780813991 (DLC) 2015042867
Series	Teen health series

Sexual health information for teens: health tips about sexual development, reproduction, contraception, and sexually transmitted infections including facts about puberty, sexuality, birth control, HIVAIDS, human papillomavirus, chlamydia, gonorrhea, herpes, and more.

LCCN	2018031481
Type of material	Book
Main title	Sexual health information for teens: health tips about sexual development, reproduction, contraception, and sexually transmitted infections including facts about puberty, sexuality, birth control, HIV/AIDS, human papillomavirus, chlamydia, gonorrhea, herpes, and more.
Edition	Fifth edition.
Published/Produced	Detroit, MI: Omnigraphics, Inc., [2018]
Description	1 online resource.
ISBN	9780780816459 (eBook)
LC classification	RA777
Related names	Omnigraphics, Inc., issuing body.
Summary	"Offers basic consumer health information for teens about puberty, development, and sexuality, maintaining sexual health, and preventing pregnancy and sexually transmitted diseases. Includes index, resource information, and online access"-- Provided by publisher.

Subjects	Teenagers--Health and hygiene. Sexual health. Reproductive health. Puberty. Sexually transmitted diseases--Prevention. Sex instruction for teenagers.
Notes	Includes bibliographical references and index. Grade 9 to 12.
Additional formats	Print version: Sexual health information for teens Fifth edition. Detroit, MI: Omnigraphics, Inc., [2018] 9780780816442 (DLC) 2018029168
Series	Teen health series

Sexual health information for teens: health tips about sexual development, reproduction, contraception, and sexually transmitted infections including facts about puberty, sexuality, birth control, HIVAIDS, human papillomavirus, chlamydia, gonorrhea, herpes, and more.

LCCN	2018029168
Type of material	Book
Main title	Sexual health information for teens: health tips about sexual development, reproduction, contraception, and sexually transmitted infections including facts about puberty, sexuality, birth control, HIV/AIDS, human papillomavirus, chlamydia, gonorrhea, herpes, and more.
Edition	Fifth edition.
Published/Produced	Detroit, MI: Omnigraphics, Inc., [2018] ©2018
Description	xii, 382 pages: illustrations; 25 cm.
ISBN	9780780816442 (hardcover: alk. paper) 0780816447 (hardcover: alk. paper)
LC classification	RA777 .S47 2018
Related names	Omnigraphics, Inc., issuing body.

Summary	"Offers basic consumer health information for teens about puberty, development, and sexuality, maintaining sexual health, and preventing pregnancy and sexually transmitted diseases. Includes index, resource information, and online access"-- Provided by publisher.
Subjects	Teenagers--Health and hygiene. Sexual health. Reproductive health. Puberty. Sexually transmitted diseases--Prevention. Sex instruction for teenagers.
Notes	Includes bibliographical references and index. Grade 9 to 12.
Additional formats	Online version: Sexual health information for teens Fifth edition. Detroit, MI: Omnigraphics, Inc., [2018] 9780780816459 (DLC) 2018031481
Series	Teen health series

Sexually transmitted diseases: a practical guide for primary care

LCCN	2013939596
Type of material	Book
Main title	Sexually transmitted diseases: a practical guide for primary care / Neil S. skolnik, Amy Lynn Clouse, Jo Ann Woodward, editors.
Edition	Second edition.
Published/Produced	New York: Humana Press, [2013]
Description	xiv, 185 pages: illustrations; 24 cm.
ISBN	9781627034982 (alk. paper) 1627034986 (alk. paper)
LC classification	RC200 .S4965 2013
Related names	Skolnik, Neil S., editor of compilation. Clouse, Amy L., editor of compilation. Woodward, JoAnn, editor of compilation.

Contents	Human Papillomavirus and Genital Warts / Linda Caryn Goldman, Amy Lynn Clouse -- Genital Herpes / D. Gene Parks -- Chlamydia / Albert John Phillips -- Gonorrhea / Jennifer E. Thuener, Amy Lynn Clouse -- Pelvic Inflammatory Disease / Joshua H. Barash, Christina Hillson -- Viral Hepatitis / Justin A. Reynolds, Jeremy Herman -- HIV: Identification, Diagnosis, and Prevention / Gunter Rieg -- Syphilis / Carolyn M. Sutton -- Trichomoniasis / Anne Moore -- Lymphogranuloma Venereum, Chancroid, Granuloma Inguinale, and Molluscum Contagiosum / Jo Ann Woodward -- Ectoparasites: Scabies and Pediculosis Pubis / Jo Ann Woodward -- Public Health and Prevention / Elissa Meites -- Evaluation and Treatment of STD Exposure and Sexual Assault / Christine M. Stroka, Mathew M. Clark.
Subjects	Sexually transmitted diseases. Sexually Transmitted Diseases--diagnosis. Primary Health Care. Sexually Transmitted Diseases--therapy.
Notes	Includes bibliographical references and index.
Series	Current clinical practice Current clinical practice.

Sexually transmitted diseases: vaccines, prevention and control

LCCN	2012540006
Type of material	Book
Main title	Sexually transmitted diseases: vaccines, prevention and control / edited by Lawrence Stanberry and Susan Rosenthal.
Edition	2nd ed.
Published/Created	London: Academic Press, c2013.

Description	xvi, 509 p.: ill. (some col.); 24 cm.
ISBN	9780123910592 (hbk.) 0123910595 (hbk.)
LC classification	RC200 S53 2013
Related names	Stanberry, Lawrence R. Rosenthal, Susan N.
Contents	Global epidemiology of sexually transmitted diseases / Adrian Mindel ... [et al.] -- The genital tract: anatomical developmental, and microbiological factors affecting sexually transmitted disease acquisition / Morris D. Cooper and Anne M. Rompalo -- Mucosal immunity in the human reproductive tract: balancing reproductive potential with protection against sexually transmitted pathogens / Charles R. Wira, Marta Rodríguez-García, and Mickey V. Patel -- Contextual factors associated with sexually transmitted infections / Marina Catallozzi, Beth A. Auslander, and Susan L. Rosenthal -- Abstinence and curricular approaches to sexually transmitted infection prevention for adolescents / Mary A. Ott -- Male circumcision: a tool for prevention of sexually transmitted diseases / Aaron A.R. Tobian and Thomas C. Quinn -- Issues in men's reproductive health / Patricia Kissinger, Akinlabi Sanusi, and David L. Bell -- Physical barrier methods: acceptance, use, and effectiveness / Adrian Mindel and Shailendra Sawleshwakar -- Topical microbicides for human immunodeficiency virus and sexually transmitted disease prevention / Kristine Torjesen ... [et al.] -- Topical microbicide acceptability / Kathleen M. Morrow, Elizabeth E. Tolley, and Alex Carballo-Dieguez -- STD vaccine acceptability in sexually

transmitted diseases / Nathan W. Stupiansky and Gregory D. Zimet -- Genital and perinatal Herpes simplex virus infections / Lawrence R. Stanberry -- The epidemiology and control of human Papillomavirus infection and clinical disease / Alan G. Nyitray ... [et al.] -- Hepatitis B virus infection / Jane N. Zuckerman and Arie J. Zuckerman -- Chlamydia / Louise M. Hafner and Peter Timms -- Gonorrhea: epidemiology, control, and prevention / Jonathan M. Zenilman and Carolyn D. Deal -- Syphilis / Rick Varma, Claudia Estcourt, and Adrian Mindel -- Bacterial vaginosis / Jeanne M. Marrazzo and Sharon L. Hillier.

Subjects	Sexually transmitted diseases--Prevention. Sexually Transmitted Diseases--epidemiology. Sexually Transmitted Diseases--prevention & control.
Notes	Includes bibliographical references and index.

Sexually transmitted diseases sourcebook: basic consumer health information about sexual health and the screening, diagnosis, treatment, and prevention of common sexually transmitted diseases (STDs), including chancroid, chlamydia, gonorrhea, herpes, hepatitis, human immunodeficiency virus/acquired immunodeficiency syndrome (HIV/AIDS), human papillomavirus (HPV), syphilis, and trichomoniasis; along with facts about risk factors and complications, trends and disparities in infection rates, tips for discussing STDs with sexual partners, a glossary of related terms, and resources for additional help and information

LCCN	2012030809
Type of material	Book
Main title	Sexually transmitted diseases sourcebook: basic consumer health information about sexual health

	and the screening, diagnosis, treatment, and prevention of common sexually transmitted diseases (STDs), including chancroid, chlamydia, gonorrhea, herpes, hepatitis, human immunodeficiency virus/acquired immunodeficiency syndrome (HIV/AIDS), human papillomavirus (HPV), syphilis, and trichomoniasis; along with facts about risk factors and complications, trends and disparities in infection rates, tips for discussing STDs with sexual partners, a glossary of related terms, and resources for additional help and information / edited by Amy L. Sutton.
Edition	5th ed.
Published/Created	Detroit, MI: Omnigraphics, Inc., c2013.
Description	xviii, 580 p.; 24 cm.
ISBN	9780780812819 (hardcover: alk. paper)
LC classification	RC200.2 .S387 2013
Related names	Sutton, Amy L.
Summary	"Provides basic consumer health information about risk factors, symptoms, testing, and treatment of sexually transmitted diseases and related complications, along with facts about prevention strategies. Includes index, glossary of related terms, and other resources"-- Provided by publisher.
Subjects	Sexually transmitted diseases--Popular works.
Notes	Includes bibliographical references and index.
Series	Health reference series

Sexually transmitted diseases sourcebook: basic consumer health information about sexual health and the screening, diagnosis, treatment, and prevention of common sexually transmitted diseases (STDs), including chancroid, chlamydia, gonorrhea, herpes, hepatitis, human immunodeficiency virus/acquired immunodeficiency syndrome (HIVAIDS), human papillomavirus (HPV), syphilis, and trichomoniasis; along with facts about risk factors and complications, trends and disparities in infection rates, tips for discussing stds with sexual partners, a glossary of related terms, and resources for additional help and information.

LCCN	2016011521
Type of material	Book
Main title	Sexually transmitted diseases sourcebook: basic consumer health information about sexual health and the screening, diagnosis, treatment, and prevention of common sexually transmitted diseases (STDs), including chancroid, chlamydia, gonorrhea, herpes, hepatitis, human immunodeficiency virus/acquired immunodeficiency syndrome (HIV/AIDS), human papillomavirus (HPV), syphilis, and trichomoniasis; along with facts about risk factors and complications, trends and disparities in infection rates, tips for discussing stds with sexual partners, a glossary of related terms, and resources for additional help and information.
Edition	Sixth edition.
Published/Produced	Detroit, MI: Omnigraphics, [2016] ©2016
Description	xvii, 657 pages: illustrations; 24 cm.
ISBN	9780780815186 (hardcover: alk. paper) 0780815181 (hardcover: alk. paper)
LC classification	RC200.2 .S387 2016

Related names	Omnigraphics, Inc.
Summary	"Provides basic consumer health information about types of STDs as well as symptoms, diagnosis, treatment, risks and prevention, and living with sexually transmitted infections Includes index, glossary of related terms, and other resources"-- Provided by publisher.
Subjects	Sexually transmitted diseases--Popular works.
Notes	Includes bibliographical references and index.
Additional formats	Online version: Sexually transmitted diseases sourcebook Sixth edition. Detroit, MI: Omnigraphics, [2016] 9780780815193 (DLC) 2016014568
Series	Health reference series

Sexually transmitted diseases sourcebook: basic consumer health information about sexual health and the screening, diagnosis, treatment, and prevention of common sexually transmitted diseases (STDs), including chancroid, chlamydia, gonorrhea, herpes, hepatitis, human immunodeficiency virus/acquired immunodeficiency syndrome (HIVAIDS), human papillomavirus (HPV), syphilis, and trichomoniasis; along with facts about risk factors and complications, trends and disparities in infection rates, tips for discussing stds with sexual partners, a glossary of related terms, and resources for additional help and information.

LCCN	2018061734
Type of material	Book
Main title	Sexually transmitted diseases sourcebook: basic consumer health information about sexual health and the screening, diagnosis, treatment, and prevention of common sexually transmitted diseases (STDs), including chancroid, chlamydia, gonorrhea, herpes, hepatitis, human

	immunodeficiency virus/acquired immunodeficiency syndrome (HIV/AIDS), human papillomavirus (HPV), syphilis, and trichomoniasis; along with facts about risk factors and complications, trends and disparities in infection rates, tips for discussing stds with sexual partners, a glossary of related terms, and resources for additional help and information.
Edition	Seventh edition.
Published/Produced	Detroit, MI: Omnigraphics, [2019]
ISBN	9780780816916 (hard cover: alk. paper)
LC classification	RC200.2 .S387 2019
Related names	Omnigraphics, Inc., issuing body.
Summary	"Provides basic consumer health information about risk factors, symptoms, testing, and treatment of sexually transmitted diseases and related complications, along with facts about prevention strategies. Includes index, glossary of related terms, and other resources"-- Provided by publisher.
Subjects	Sexually transmitted diseases--Popular works.
Notes	Includes index.
Additional formats	Online version: Sexually transmitted diseases sourcebook Seventh edition. Detroit, MI: Omnigraphics, [2019] 9780780816923 (DLC) 2019000435
Series	Health reference series

Sexually transmitted diseases sourcebook: basic consumer health information about sexual health and the screening, diagnosis, treatment, and prevention of common sexually transmitted diseases (STDs), including chancroid, chlamydia, gonorrhea, herpes, hepatitis, human immunodeficiency virus/acquired immunodeficiency syndrome (HIVAIDS), human papillomavirus (HPV), syphilis, and trichomoniasis; along with facts about risk factors and complications, trends and disparities in infection rates, tips for discussing stds with sexual partners, a glossary of related terms, and resources for additional help and information.

LCCN	2019000435
Type of material	Book
Main title	Sexually transmitted diseases sourcebook: basic consumer health information about sexual health and the screening, diagnosis, treatment, and prevention of common sexually transmitted diseases (STDs), including chancroid, chlamydia, gonorrhea, herpes, hepatitis, human immunodeficiency virus/acquired immunodeficiency syndrome (HIV/AIDS), human papillomavirus (HPV), syphilis, and trichomoniasis; along with facts about risk factors and complications, trends and disparities in infection rates, tips for discussing stds with sexual partners, a glossary of related terms, and resources for additional help and information.
Edition	Seventh edition.
Published/Produced	Detroit, MI: Omnigraphics, [2019]
Description	1 online resource.
ISBN	9780780816923 (ebook)
LC classification	RC200.2
Related names	Omnigraphics, Inc., issuing body.
Summary	"Provides basic consumer health information about risk factors, symptoms, testing, and

	treatment of sexually transmitted diseases and related complications, along with facts about prevention strategies. Includes index, glossary of related terms, and other resources"-- Provided by publisher.
Subjects	Sexually transmitted diseases--Popular works.
Notes	Includes index.
Additional formats	Print version: Sexually transmitted diseases sourcebook Seventh edition. Detroit, MI: Omnigraphics, [2019] 9780780816916 (DLC) 2018061734
Series	Health reference series

Small-molecule inhibitors of protein-protein interactions

LCCN	2013444357
Type of material	Book
Main title	Small-molecule inhibitors of protein-protein interactions / Lyubomir Vassilev, David Fry, editors.
Published/Created	Heidelberg; New York: Springer Verlag, c2011.
Description	x, 174 p.: ill. (some col.); 24 cm.
Links	http://d-nb.info/1009722301/04 Inhaltsverzeichnis
ISBN	9783642170829 (alk. paper) 364217082X (alk. paper) 9783642170836 (e-ISBN) 3642170838 (e-ISBN)
LC classification	QP551.5 .S63 2011 QR1 .E6 v.348
Related names	Vassilev, Lyubomir. Fry, David, Dr.
Contents	Hydrogen-bonded synthetic mimics of protein secondary structure as disruptors of protein-protein interactions / M.J. Adler, A.G. Jamieson

and Andrew D. Hamilton -- Small-molecule inhibitors of IL-2/IL-2R: Lessons learned and applied / C.G. Wilson and M.R. Arkin -- Small molecule inhibitors of the human papillomavirus E1-E2 interaction / P.W. White, Anne-Marie Faucher and Nathalie Goudreau -- Design of small-molecule Smac mimetics as IAP antagonists / Shaomeng Wang -- Small-molecule inhibitors reveal a new function for Bcl-2 as a proangiogenic signaling molecule / Benjamin D. Zeitlin and Jacques E. Nör -- Small-molecule modulators of c-Myc/Max and Max/Max interactions / Thorsten Berg -- Small-molecule inhibitors of the p53-MDM2 interaction / Binh T. Vu and Lyubomir Vassilev.

Subjects Protein-protein interactions.
Enzyme inhibitors.
Protein Binding.
Drug Design.
Small Molecule Libraries.
Enzyme inhibitors.
Protein-protein interactions.
Protein-Protein-Wechselwirkung.
Inhibitor.
Niedermolekulare Verbindungen.

Form/Genre Aufsatzsammlung.

Notes Includes bibliographical references and index.

Series Current topics in microbiology and immunology, 0070-217X; v. 348
Current topics in microbiology and immunology; v. 348. 0070-217X

The dental hygienist's guide to nutritional care

LCCN	2013042535
Type of material	Book
Personal name	Stegeman, Cynthia A., author.
Main title	The dental hygienist's guide to nutritional care / Cynthia A. Stegeman, RDH, EdD, RDN, LD, CDE, Ohio Delegate to the Academy of Nutrition and Dietetics, Associate Professor, Dental Hygiene Program, University of Cincinnati, Cincinnati, Ohio, Judith Ratliff Davis, MS, RDN, Former Quality Assurance Nutrition Consultant, Women, Infants and Children (WIC) Program, Texas Department of State Health Services, Austin, Texas.
Edition	Edition 4.
Published/Produced	St. Louis, Missouri: Elsevier Saunders, [2015] ©2015
Description	xvii, 450 pages: illustrations (chiefly color); 28 cm
ISBN	9781455737659 (pbk.: alk. paper) 1455737658 (pbk.: alk. paper)
LC classification	RK60.7 .S74 2015
Related names	Davis, Judi Ratliff, author.
Contents	Pt. I. Orientation to basic nutrition. 1. Overview of healthy eating habits: Basic nutrition; Physiological functions of nutrients; Basic concepts of nutrition; Government nutrition concerns; Nutrient recommendations: dietary reference intakes; Food guidance systems for Americans; Dietary guidelines for Americans; Myplate system; Other food guides; Nutrition labeling; Health application 1: Obesity -- 2. Concepts in biochemistry: What is biochemistry?; Fundamentals of biochemistry; Principle

biomolecules in nutrition; Summary of metabolism; Health application 2: Lactose intolerance -- 3. The alimentary canal: digestion and absorption: Physiology of the gastrointestinal tract; Oral cavity; Esophagus; Gastric digestion; Small intestine; Large intestine; Health application 3: Gluten-related disorders -- 4. Carbohydrate: the efficient fuel: Classification; Physiological roles; Requirements; Sources; Hyperstates and hypostates; Nonnutritive sweeteners/sugar substitutes; Health application 4: High-fructose corn syrup -- 5. Protein: the cellular foundation: Amino acids; Classification; Physiological roles; Requirements; Sources; Underconsumption and health-related problems; Overconsumption and health-related problems; Health application 5: Vegetarianism -- 6. Lipids: the condensed energy: Classification; Chemical structure; Characteristics of fatty acids; Compound lipids; Cholesterol; Physiological roles; Dietary fats and dental health; Dietary requirements; Sources; Overconsumption and health-related problems; Underconsumption and health-related problems; Fat replacers; Health application 6: Hyperlipidemia -- 7. Use of the energy nutrients: metabolism and balance: Metabolism; Role of the liver; Role of the kidneys; Carbohydrate metabolism; Protein metabolism; Lipid metabolism; Alcohol metabolism; Metabolic interrelationships; Metabolic energy; Basal metabolis rate; Total energy requirements; Energy balance; Inadequate energy intake; Health application 7: Diabetes Mellitus -- 8. Vitamins required for calcified structures: Overview of vitamins; Vitamin A (retinol, carotene); Vitamin D (calciferol); Vitamin E (tocopherol); Vitamin K; Vitamin C (ascorbic acid); Health application 8: Antioxidants -- 9. Minerals essential for calcified structures: Bone mineralization and growth; Formation of teeth; Introduction to minerals; Calcium; Phosphorus; Magnesium; Fluoride; Health application 9: Osteoporosis -- 10. Nutrients present in calcified structures: Copper; Selenium; Chromium; Manganese; Molybdenum; Ultratrace elements; Health application 10: Alzheimer disease -- 11. Vitamins required for oral soft tissues and salivary glands: Physiology of soft tissues; Thiamin (vitamin B&#27;b1&#27;s); Riboflavin (vitamin B&#27;b2&#27;s); Niacin (vitamin B&#27;b3&#27;s); Pantothenic acid (vitamin B&#27;b5&#27;s); Vitamin B&#27;b6&#27;s (pyridoxine); Folate/folic acid; Vitamin B&#27;b12&#27;s (cobalamin); Biotine (vitamin B&#27;b7&#27;s); Other vitamins; Health application 11: Vitamin, mineral, and herbal supplements -- 12. Fluids and minerals required for oral soft tissues and salivary glands: Fluids; Electrolytes; Sodium; Chloride; Potassium; Iron; Zinc; Iodine; Health application 12: Hypertension --

Pt. II. Application of nutrition principles. 13. Nutritional requirements affecting oral health in women: Health pregnancy; Lactation; Oral contraceptive agents; Menopause; Health application 13: Fetal alcohol spectrum disorder -- 14. Nutritional requirements during growth and development and eating habits affecting oral health: Infants; Dietary recommendations and guidelines for growth (children older than 2 years of age); Toddler and preschool children; Attention-deficit/hyperactivity disorders; Children with special needs; School-age children (7 to 12 years old); Adolescents; Health application 14: Childhood and adolescent obesity -- 15. Nutritional requirements for older adults and eating habits affecting oral health: General health status; Physiological factors influencing nutritional needs and status; Socioeconomic and psychological factors; Nutrient requirements; Eating patterns; MyPlate for older adults; Health application 15: Genomics -- 16. Food factors affecting health: Healthcare disparities; Food patterns; Working with patients with different food patterns; Food budgets; Maintaining optimal nutrition during food preparation; Food fads and misinformation; Referrals for nutritional resources; Role of dental hygienists; Health application 16: Food insecurity in the United States -- 17. Effects of systemic disease on nutritional status and oral health: Effects of chronic disease on intake; Anemias; Other hematological disorders; Gastrointestinal problems; Cardiovascular conditions; Skeletal system; Metabolic problems; Neuromuscular problems; Neoplasia; Acquired immunodeficiency disease; Mental health problems; Health application 17: Human papillomavirus --

pt. III. Nutritional aspects of oral health. 18. Nutritional aspects of dental caries: causes,

	prevention, and treatment: Major factors in the dental caries process; Other factors influencing carcinogenicity; Dental plan; Health application 18: Genetically modified foods -- 19. Nutritional aspects of gingivitis and periodontal disease: Physical effects of food on periodontal health; Nutritional considerations for periodontal patients; Gingivitis; Chronic periodontitis; Necrotizing periodontal disease; Health application 19: Tobacco cessation -- 20. Nutritional aspects of alterations in the oral cavity: Orthodontics; Xerostomia; Root caries and dentin hypersensitivity; Dentition status; Oral and maxillofacial surgery; Loss of alveolar bone; Glossitis; Temporomandibular disorder; Health application 20: Functional foods -- 21. Nutritional assessment and education for dental patients: Evaluation of the patient; Assessment of nutritional status; Identification of nutritional status; Formation of nutrition treatment plan; Facilitative communication skills; Health application 21: Health literacy.
Subjects	Nutrition and dental health. Dental hygienists. Nutritional Requirements. Oral Health. Dental Hygienists.
Notes	Includes bibliographical references and index.

The new public health

LCCN	2014427175
Type of material	Book
Personal name	Tulchinsky, Theodore H., author.

Main title	The new public health / Theodore H. Tulchinsky, MD, MPH, Braun School of Public Health, Hebrew University-Hadassah, Ein Karem, Jerusalem, Israel and Elena A. Varavikova, MD, MPH, PhD, Research Institute for Public Health Organization and Information (CNIIOIZ), Moscow, Russian Federation with Joan D. Bickford, MSN, Former Chief Public Health Nurse, Province of Manitoba, Canada; foreword by Jonathan Fielding, MD, MPH, University of Los Angeles (UCLA), California.
Edition	Third edition.
Published/Produced	Amsterdam; Boston: Elsevier/AP, [2014] ©2014
Description	xxv, 884 pages: illustrations; 28 cm
ISBN	9780124157668 (hbk) 0124157661 (hbk)
LC classification	RA425 .T77 2014
Related names	Varavikova, Elena, author. Bickford, Joan D., author.
Summary	The book provides a unified approach to public health appropriate for all masters' level students and practitioners-specifically for courses in MPH programs, community health and preventive medicine programs, community health education programs, community health nursing programs.
Contents	1. A History of Public Health -- Introduction -- Prehistoric Societies -- The Ancient World -- The Early Medieval Period (Fifth to Tenth Centuries CE) -- The Late Medieval Period (Eleventh to Fifteenth Centuries) -- The Renaissance (1400--1600s) -- Enlightenment, Science, and Revolution (1600s--1800s) -- Eighteenth-Century Reforms -- Applied Epidemiology -- Jenner and Vaccination

-- Foundations of Health Statistics and Epidemiology -- Social Reform and the Sanitary Movement (1830--1875) -- Social Security -- Snow on Cholera -- The Germ and Miasma Theories -- Hospital Reform -- The Bacteriological Revolution -- Pasteur, Cohn, Koch, and Lister -- The Basis of Genetics -- Vectorborne Disease -- Microbiology and Immunology -- Poliomyelitis -- Advances in Prevention and Treatment of Infectious Diseases -- Infections and Chronic Diseases -- Maternal and Child Health -- Nutrition in Public Health -- Military Medicine -- Globalization of Health -- Epidemiological Transition -- Achievements of Public Health in the Twentieth Century -- Creating and Managing Health Systems -- Challenges in the Twenty-First Century -- Summary -- Historical Markers -- Note -- Bibliography -- 2.Expanding the Concept of Public Health -- Introduction -- Concepts of Public Health -- Evolution of Public Health -- Health and Disease -- The Host-Agent--Environment Paradigm -- The Expanded Host-Agent-Environment Paradigm -- The Natural History of Disease -- Society and Health -- Health Insurance Systems -- Health Promotion -- Social Determinants of Health -- Modes of Prevention -- Health Promotion -- Health Protection -- Primary Prevention -- Secondary Prevention -- Tertiary Prevention -- Demographic and Epidemiological Transition -- Interdependence of Health Services -- Defining Public Health -- Social Medicine and Community Health -- Social Hygiene, Eugenics, and Corruption of Public Health Concepts --

Medical Ecology -- Community-Oriented Primary Care -- The World Health Organization's Definition of Health -- Alma-Ata: Health for All -- Selective Primary Care -- The Risk Approach -- The Case for Action -- Political Economy and Health -- Social, Behavioral, and Political Sciences -- Health and Development -- Health Systems: The Case for Reform -- Advocacy and Consumerism -- Professional Advocacy and Resistance -- Consumerism -- The Health Field Concept -- The Value of Medical Care in Public Health -- Health Targets -- US Health Targets -- European Health Targets -- UK Health Targets -- Individual and Community Participation in Health -- Founding of Health Promotion: The Ottawa Charter of 1986 -- State and Community Models of Health Promotion -- Healthy Cities/Towns/Municipalities -- Human Ecology and Health Promotion -- Defining Public Health Standards -- Integrative Approaches to Public Health -- Achievements of Public Health -- The Future of Public Health -- The New Public Health -- Summary -- Note -- Bibliography -- 3.Measuring, Monitoring, and Evaluating the Health of a Population -- Introduction -- Demography -- Fertility -- Population Pyramid -- Life Expectancy -- Epidemiology -- Social Epidemiology -- Epidemiology in Building Health Policy -- Definitions and Methods of Epidemiology -- Rates and Ratios -- Measures of Disease Occurrence or Morbidity -- Prevalence Rates -- Measures of Mortality -- Social Classification -- Sentinel Events -- The Burden of Disease -- Years of Potential Life Lost --

Qualitative Measures of Morbidity and Mortality -- Measurement -- Research and Survey Methods -- Variables -- The Null Hypothesis -- Confounders -- Sampling -- Random Sampling -- Normal Distribution -- Standardization of Rates -- Direct Method of Standardization -- Indirect Method of Standardization -- Potential Errors in Measurement -- Reliability -- Validity -- Screening for Disease -- Epidemiological Studies -- Observational Studies -- Experimental Epidemiology -- Establishing Causal Relationships -- Notification of Diseases -- Special Registries and Reporting Systems -- Disease Classification -- Hospital Discharge Information -- Health Information Systems (Informatics) -- WHO European Region Health for All Database -- Surveillance, Reporting, and Publication -- Assessing the Health of the Individual -- Assessment of Population Health -- Defining the Population -- Health Care Financing and Organization -- Health Care Resources -- Utilization of Services -- Health Care Outcomes -- Quality of Care -- Self-Assessment of Health -- Costs and Benefits -- Effects of Intervention -- Qualitative and Quantitative Research Methods -- From Health Information to Knowledge to Policy -- Summary -- Note -- Bibliography -- 4.Communicable Diseases -- Introduction -- Public Health and the Control of Communicable Disease -- The Nature of Communicable Disease -- Host-Agent-Environment Triad -- Classifications of Communicable Diseases -- Modes of Transmission of Disease -- Immunity -- Surveillance -- Healthcare-Associated Infections -

- Endemic and Epidemic Disease -- Epidemic Investigation -- Control of Communicable Diseases -- Treatment -- Methods of Prevention -- Vaccine-Preventable Diseases -- Immunization Coverage -- Vaccine Regulation, Schedules, and Programs -- Essentials of an Immunization Program -- Regulation of Vaccines -- Vaccine Development -- Connections Between Infectious and Chronic Diseases -- Control and Eradication of Infectious Diseases -- Success in Eradicating Smallpox -- Eradication of Poliomyelitis -- Future Candidates for Disease -- Eradication -- Criteria for Eradicable Diseases -- Directly Observed Therapy Short Course (DOTS) -- Vectorborne Diseases -- Malaria -- Rickettsial Infections -- Arboviruses (Arthropod-Borne Viral Diseases) -- Lyme Disease -- Neglected Tropical Diseases -- Echinococcosis -- Tapeworm -- Onchocerciasis -- Dracunculiasis -- Schistosomiasis -- Leishmaniasis -- Trypanosomiasis -- Other Parasitic Diseases -- Legionnaire's Disease -- Leprosy -- Sexually Transmitted Infections -- Syphilis -- Gonorrhea -- Chancroid -- Herpes Simplex -- Chlamydia -- Trachoma -- Trichomoniasis -- Human Papillomavirus -- Control of Sexually Transmitted Infections -- HIV/AIDS -- Diarrheal Diseases -- Salmonella -- Shigella -- Escherichia coli -- Cholera -- Viral Gastroenteritis -- Parasitic Gastroenteritis -- A Program Approach to Diarrheal Disease Control -- Acute Respiratory Infections -- Inequalities in Control of Communicable Diseases -- Communicable Disease Control in the New Public Health -- Summary -- Note -- Bibliography

-- 5.Non-Communicable Diseases and Conditions -- Introduction -- The Rise and Spread of Chronic Disease -- Environment and Lifestyle -- Global Burden of Mortality -- Risk Factors, Health Promotion, and Prevention of Non-Communicable Diseases -- Tobacco -- Alcohol and Other Drugs -- Physical Activity -- Hypertension -- Diet and Obesity -- Mental Illness -- Genetic, Infectious, and Non-Communicable Diseases -- The Social and Economic Burden of Non-Communicable Diseases -- Health Promotion and Disease Prevention -- Hypertension and Cardiovascular Diseases -- Diabetes Mellitus -- Prevention of Diabetes and Its Complications -- Cancer -- Prevention of Carton -- Trauma, Violence, and Injury -- Motor Vehicle Accidents -- Domestic and Sexual Violence -- Suicide -- Poisoning -- Homicide -- Prevention of Violence -- Chronic Lung Disease -- Asthma -- Chronic Obstructive Pulmonary Disease -- Restrictive Lung Diseases -- Occupational Lung Diseases -- End-Stage Renal Disease -- Disabling Conditions -- Arthritis and Musculoskeletal Disorders -- Osteoporosis -- Degenerative Osteoarthritis -- Rheumatoid Arthritis and Gout -- Low Back and Neck Pain Syndromes -- Neurological Disorders -- Alzheimer's Disease -- Parkinson's Disease -- Multiple Sclerosis -- Epilepsy or Seizures -- Brain and Spinal Cord Injury -- Visual Disorders -- Hearing Disorders -- Non-Communicable Diseases and the New Public Health -- Integrated Non-Communicable Disease Prevention and Control -- Summary -- Note -- Bibliography --

Recommended Reading -- Global Burden of Non-Communicable -- Diseases -- Social Determinants, Environment, and Economics of Non-Communicable Diseases -- Risk Factors, Health Promotion, and Prevention of Non-Communicable Diseases -- Smoking -- Diet, Physical Activity, and Obesity -- Infectious Diseases and Non-Communicable Diseases -- Hypertension and Cardiovascular Diseases -- Metabolic Syndrome, Diabetes Mellitus, and End-Stage Renal Disease -- Cancer -- Trauma, Violence, Injury, and Suicide -- Chronic Lung Disease -- Disabling Conditions -- 6.Family Health -- Introduction -- The Family Unit -- Preventive Health through Life Stages -- The Family Physician Cannot do All This Alone -- Maternal Health -- Fertility -- Public Health Challenges of Fertility -- Family Planning -- Maternal Mortality and Morbidity -- Pregnancy Care -- High-Risk Pregnancy -- Labor and Delivery -- Safe Motherhood Initiatives -- Care of the Newborn -- Low Birth Weight/Preterm Birth -- Postdelivery Care (Puerperium) -- Genetic and Birth Disorders -- Rhesus Hemolytic Disease of the Newborn -- Neural Tube Defects -- Cerebral Palsy -- Low Birth Weight and Developmental Disability -- Down Syndrome -- Cystic Fibrosis -- Sickle-Cell Disease -- Thalassemia -- Phenylketonuria -- Congenital Hypothyroidism -- Congenital Syphilis -- Fetal Alcohol Syndrome -- Tay--Sachs Disease -- Glucose-6-Phosphate Dehydrogenase Deficiency -- Familial Mediterranean Fever -- Infant and Child Health -- Fetal and Infant Mortality -- Infancy Care and

Feeding -- Anticipatory Counseling -- Documentation, Records, and Monitoring -- Preschool Children (Age 1--5 Years) -- School and Adolescent Health -- Smoking -- Alcohol Abuse -- Drug Abuse -- Sexual Risk Behavior -- Dietary Risk Behavior -- Physical Activity -- Violence and Gang Behavior -- Adult Health -- Women's Health -- Men's Health -- Health of Older Adults -- Health Maintenance for Older Adults -- Summary -- Note -- Bibliography -- 7.Special Community Health Needs -- Introduction -- Mental Health -- Historical Changes in Methods of Treatment -- Mental Health Epidemiology -- Mental Disorder Syndromes -- Controversies in Mental Health Policies -- Community-Oriented Mental Health -- Prevention and Health Promotion -- Intellectual Disability -- Oral Health -- Fluoridation -- Periodontal Disease -- Dental Care -- Oral Cancer -- Physical Disability and Rehabilitation -- Special Croup Health Needs -- Cay and Lesbian Health -- Native Peoples' Health -- Prisoners' Health -- Migrant Population Health -- Homeless Population Health -- Refugee Health -- Military Medicine -- Health in Disasters -- Summary -- Note -- Bibliography -- Mental Health -- Oral Health -- Disability and Rehabilitation -- Cay and Lesbian Health -- Native Peoples' Health -- Prisoners' Health -- Migrant and Refugee Health -- Military Medicine -- Disasters and Health -- 8.Nutrition and Food Safety -- Introduction -- Development of Nutrition in Public Health -- Nutrition in a Global Context -- Nutrition and Infection -- Functions of Food -- Human

Nutritional Requirements -- Carbohydrates -- Proteins -- Fats and Oils -- Vitamins -- Minerals and Trace Elements -- Growth -- Measuring Body Mass -- Recommended Dietary Intakes -- Disorders of Undernutrition -- Underweight: Protein---Energy Malnutrition -- Failure to Thrive -- Marasmus -- Kwashiorkor -- Micronutrient Deficiencies -- Osteoporosis -- Eating Disorders -- Diseases of Overnutrition -- Overweight/Obesity -- Diabetes Mellitus -- Cardiovascular Diseases -- Cancer -- Nutrition in Pregnancy and Lactation -- Promoting Healthy Diets and Lifestyles -- Dietary Guidelines -- Vitamin and Mineral Enrichment of Basic Foods -- Controversy in Food Enrichment -- Genetically Modified Foods -- Food and Nutrition Policy -- Evolution of a Federal Role -- Nutrition Issues in Development Policies -- Role of the Private Sector and Non-Governmental Organizations -- Role of Health Providers -- Nutrition Monitoring and Evaluation -- Standard Reference Populations -- Measuring Deviation from the Reference Population -- Global Nutrition Targets for 2025 -- Food Quality and Safety -- Nutrition and the New Public Health -- Summary -- Note -- Bibliography -- Recommended Reading -- Foodborne Illness, Nutrition, and Infectious Disease -- Monitoring Growth and Body Mass -- Nutrient Requirements, Recommended Dietary Intakes, and Dietary Guidelines -- Micronutrient Deficiency Conditions -- Vitamin D and Bone Health -- Nutrition, Obesity, and Chronic Disease -- Maternal/Child Nutrition and Survival -- Food Fortification -- Food and Nutrition Policy, and Food Security -- Nutrition Monitoring and

Evaluation -- 9.Environmental and Occupational Health -- Environmental Health -- Introduction -- Global Environmental Issues -- Geographic and Environmental Epidemiology -- Environmental Targets -- Environment and Infectious Diseases -- Global Climate Change -- Environmental Impact on Health Burden of Disease -- Community Water Supplies -- Waterborne Diseases -- Waterborne Disease Surveillance and Prevalence, USA -- Water as a Global Resource -- Sewage Collection and Treatment -- Solid Waste -- Toxins -- Toxic Effects on Fertility -- Toxic Effects of Lead in the Environment -- Agricultural and Environmental Hazards -- Air Pollution -- The External Environment -- Methyl Tertiary Butyl Ether -- Indoor Pollution -- Effects of Indoor Air Pollution -- Radon Gas -- Outdoor--Indoor Pollutants -- Biological Pollutants -- Built Environment and Health -- Hazardous or Toxic Wastes -- Minimata Disease -- Toxic Waste Management -- Radiation -- Ionizing Radiation -- Non-Ionizing Radiation -- Environmental Impact -- Emergency Events Involving Hazardous Substances -- Human-Caused Disasters -- War, Terrorism, and Genocide -- Preventing and Managing Environmental Emergencies -- Environmental Health Organization -- Occupational Health -- Introduction -- Development of Occupational Health -- The Health of Workers -- The Burden of Occupational Morbidity and Mortality -- Occupational Health Priorities in the USA -- International Issues in Occupational Health -- National and Management Responsibilities -- Standards and Monitoring -- Occupational Health

Targets -- Occupational Injury -- Toxicity at the Workplace and in the Environment -- Lead -- Asbestos -- Silica -- Cotton Dust (Byssinosis) -- Vinyl Chloride -- Agent Orange -- Workplace Violence -- Occupational Health in Clinical Practice -- Inspecting the Place of Work -- Risk Assessment -- Preventing Disasters in the Workplace -- Occupation and The New Public Health -- Summary -- Note -- Bibliography -- Environmental Health -- Occupational Health -- 10.Organization of Public Health Systems -- Introduction -- Government and Health of the Nation -- Federal and Unitary States -- Checks and Balances in Health Authority -- Government and the Individual -- Functions of Public Health -- Regulatory Functions of Public Health Agencies -- Investigative Functions of Public Health -- Prevention of Injuries -- Methods of Providing or Assuring Services: Direct or Indirect? -- Non-Governmental Roles in Health -- Disasters and Public Health Preparedness -- Medical Practice and Public Health -- Incentives and Regulation -- Promotion of Research and Teaching -- Accreditation and Quality Regulation -- National Government Public Health Services -- State Government Public Health Services -- Local Health Authorities -- Accreditation of Public Health Departments -- Monitoring Health Status -- National Health Targets -- Universal Health Coverage and the New Public Health -- Hospitals in the New Public Health -- Hospital Classification -- Supply of Hospital Beds -- The Changing Role of the Hospital -- Innovations in Health Care Delivery -- The Uninsured as a Public

Health Challenge -- US Patient Protection and Affordable Care Act (Obamacare) -- Summary -- Note -- Bibliography -- Agencies and Organizations -- Recommended Reading -- 11.Measuring Costs: The Economics of Health -- Introduction -- Economic Issues of Health Systems -- Investing in Health -- Basic Concepts in Health Economics -- Supply, Need, Demand, and Utilization of Health Services -- Normative Needs -- Felt Need -- Expressed Need -- Comparative Need -- Demand -- Supply -- Grossman's Demand Model -- Competition in Health Care -- Elasticities of Demand -- Measuring Costs -- Economic Measures of Health Status -- Cost--Benefit Analysis -- Cost-Effectiveness Analysis -- Basic Assessment Scheme for Intervention Costs and Consequences -- The Value of Human Life -- Health Financing: The Macroeconomic Level -- Costs of Illness -- Costs and Variations in Medical Practice -- Cost Containment -- Medical and Hospital Care: Microeconomics -- Payment for Doctors' Services -- Payment for Comprehensive Care -- Health Maintenance and Managed Care Organizations -- District Health Systems -- Paying for Hospital Care -- Capital Costs -- Hospital Supply, Utilization, and Costs -- Modified Market Forces -- Economics of Prevention -- Game Theory, Economics, and Health Policy -- Economics and the New Public Health -- Summary -- Note -- Bibliography -- 12.Planning and Managing Health Systems -- Introduction -- Health Policy and Planning as Context -- The Elements of Organizations -- Scientific Management --

Bureaucratic Pyramidal Organizations -- Organizations as Energy Systems -- Cybernetics and Management -- Target-Oriented Management -- Operations Research -- Management by Objectives -- Human Relations Management -- The Hawthorne Effect -- Maslow's Hierarchy of Needs -- Theory X--Theory Y -- Network Organization -- Total Quality Management -- Changing Human Behavior -- Empowerment -- Strategic Management of Health Systems -- Health System Organization Models -- Functional Model -- Corporate Model -- Matrix Model -- Skills for Management -- The Chief Executive Officer of Health Organizations -- Community Participation -- Integration: Lateral and Vertical -- Norms and Performance Indicators -- Health Promotion and Advocacy -- Philanthropy and Volunteerism -- New Organizational Models -- New Projects and their Evaluation -- Competencies in Health Policy and Management -- Systems Approach and National Planning -- Summary -- Note -- Bibliography -- Electronic Resources -- Recommended Reading -- 13.National Health Systems -- Introduction -- Health Systems in the Industrialized Countries -- Evolution of Health Systems -- Social Insurance -- National Health Service -- National Health Insurance -- The United States -- Federal Health Initiatives -- Medicare and Medicaid -- The Changing Health Care Environment -- Health Information -- Health Targets -- Social Inequalities -- The Dilemma of the Uninsured -- Summary -- Canada -- Reform Pressures and Initiatives -- Provincial Health Reforms -- Health

Status -- Summary -- The United Kingdom -- The National Health Service -- Reforms in the 1970s and 1980s -- Reforms Since 1990 -- Social and Regional Inequalities -- Health Promotion -- Health Reforms -- Primary Care Trusts -- Summary -- The Nordic Countries -- Sweden -- Denmark -- Norway -- Finland -- Western Europe -- Germany -- The Netherlands -- Russia -- The Soviet Model -- Epidemiological Transition -- Post-Soviet Reform -- Future Prospects -- Summary -- Israel -- Origins of the Israeli Health System -- Health Resources and Expenditures -- Health Reforms -- Mental Health -- Healthy Israel 2020 -- Quality Assurance -- Summary -- Health Systems in Developing Countries -- Federal Republic of Nigeria -- Female Genital Mutilation -- Communicable Diseases -- Non-Communicable Diseases -- Nigeria's Health System -- Millennium Development Goals -- Cancer -- HIV/AIDS -- Food Fortification/Malnutrition -- Health Reform -- Summary -- Latin America and the Caribbean -- Colombia -- Asia -- India -- China -- Japan -- Comparing National Health Systems -- Economic Issues in National Health Systems -- Reforming National Health Systems -- Summary -- Note -- Bibliography -- International -- United States -- Canada -- United Kingdom -- Europe -- Nordic Countries, Germany, and the Netherlands -- Russia -- Israel -- Developing Countries -- Nigeria -- Latin America and the Caribbean -- Colombia -- India -- China -- Japan -- 14.Human Resources for Health -- Introduction -- Overview of Human Resources -- Human Resources Planning --

Supply and Demand -- Medical Education -- Postgraduate Medical Training -- Specialization and Family Practice -- Training in Preventive Medicine -- Nursing Education -- In-Service and Continuing Education -- Accreditation of Health Professional Educational or Training Facilities -- The Range of Health Disciplines -- Licensure and Supervision -- Constraints on the Health Care Provider -- Other Health Professions -- Expanding Roles in Primary Care -- Nurse Practitioners -- Physician Assistants -- Feldshers -- Community Health Workers -- Alternative Medicine -- Changing the Balance -- Education for Public Health and Health Management -- Core Elements for Master of Public Health Programs -- Ranking Universities -- Field Epidemiology -- Public Health Education in Europe -- Health Policy and Management of Human Resources -- Summary -- Note -- Bibliography -- 15.Health Technology, Quality, Law, and Ethics -- Introduction -- Innovation, Regulation, and Quality Control -- Appropriate Health Technology -- Priority Interventions in Low- and Medium-Income Countries -- Priority Selection in High-Income Countries -- Health Technology Assessment -- Technology Assessment in Hospitals -- Technology Assessment in Prevention and Health Promotion -- Technology Assessment in National Health Systems -- Dissemination of Technology -- Diffusion of Technology -- Quality Assurance -- Adverse Events and Negligence -- Licensure and Certification -- Health Facility Accreditation -- Peer Review -- Tracer Conditions -- Setting Standards -- Algorithms and Clinical Guidelines -

- Organization of Care -- Diagnosis-Related Groups -- Managed Care -- Performance Indicators -- Consumerism and Quality -- The Public Interest -- Total Quality Management -- Public Health Law -- Environmental Health -- Public Health Law Reform -- Ethical Issues In Public Health -- Individual and Community Rights -- Tragic Deviations in Public Health Ethics -- Human Experimentation -- Ethics in Public Health Research -- Ethics in Patient Care -- Sanctity of Life Versus Euthanasia -- The Imperative to Act or Not Act in Public Health -- Summary -- Note -- Bibliography -- Health Technology -- Quality -- Law -- Ethics -- 16.Global Health -- Introduction -- The Global Health Situation -- Priorities in Global Health -- Poverty-Illness-Population--Environment -- Child Health -- Maternal Health -- Population Growth -- Malnutrition -- The Fight Against HIV/AIDS and Other Communicable Diseases -- Non-Communicable Diseases -- Disaster Management -- Environment -- Global Partnership for Development -- Development and Health -- Organization for International Health -- World Health Organization -- United Nations Children's Fund -- Non-Governmental Organizations -- The World Bank -- Trends in Global Health -- Emerging Infectious Disease Threats -- Expanding National Health Capacity -- Top-Down and Bottom-Up Development -- Global Health and the New Public Health -- Summary -- Note -- Bibliography -- Global Health -- International Organizations -- Global Health in the Future.

Subjects	Public health. Medical policy.
Notes	Previous edition: London: Academic, 2009. Includes bibliographical references and index.

The pictorial atlas of common genito-urinary medicine

LCCN	2011278363
Type of material	Book
Personal name	Pareek, Shiv Shanker.
Main title	The pictorial atlas of common genito-urinary medicine / Shiv Shanker Pareek; foreword by Rob Miller.
Published/Created	London; New York: Radcliffe Pub., c2012.
Description	x, 177 p.: ill. (chiefly col.); 25 cm.
ISBN	9781846194757 (pbk.: alk. paper) 184619475X (pbk.: alk. paper)
LC classification	RC871 .P37 2012
Portion of title	Common genito-urinary medicine
Contents	Genitalia -- Syphilis -- Gonorrhoea -- Urethritis due to chlamydia trachomatis -- Non-specific urethritis -- Pelvic inflammatory disease -- Prostatitis -- Epididymitis and epididymo-orchitis -- Reiter's syndrome -- Bacterial vaginosis -- Trichomoniasis -- Genital candidiasis -- Tropical genital and sexually acquired infections -- Non-venereal endemic treponematoses -- Urinary tract infection -- Genital herpes -- Human papillomavirus -- Molluscum contagiosum -- Sexually acquired viral hepatitis B and C -- Scabies -- Pediculosis pubis -- Dermatoses affecting the genital area -- Malignant and pre-malignant conditions affecting the genital area -- Miscellaneous conditions affecting the genitalia.
Subjects	Genitourinary organs--Atlases.

	Female Urogenital Diseases--Atlases. Male Urogenital Diseases--Atlases. Urologic Diseases--Atlases.
Notes	Includes bibliographical references and index.

The women's health companion to Maternal-child nursing care: optimizing outcomes for mothers, children & families

LCCN	2015940175
Type of material	Book
Personal name	Hisley, Shelton M.
Main title	The women's health companion to Maternal-child nursing care: optimizing outcomes for mothers, children & families / Shelton M. Hisley.
Edition	Second edition.
Published/Produced	Philadelphia: F.A. Davis, [2016] ©2016
Description	ix, 193 pages: color illustrations; 28 cm
ISBN	9780803657045 (paperback) 0803657048 (paperback)
LC classification	RG951 .W375 2016
Contents	Promoting mental health and physical safety through an understanding of intimate partner violence -- Promoting reproductive health through an understanding of premenstrual syndrome -- Promoting breast health -- Promoting reproductive health through an understanding of various gynecological disorders -- Promoting reproductive health through an understanding of sexually transmitted diseases -- Promoting reproductive health through an understanding of cervical cytology screening, human papillomavirus, and cervical cancer -- Promoting menopausal health.
Subjects	Maternity nursing--Handbooks, manuals, etc.

	Pediatric nursing--Handbooks, manuals, etc.
	Maternal-Child Nursing--methods--Handbooks.
	Cultural Diversity--Handbooks.
	Evidence-Based Nursing--methods--Handbooks.
	Holistic Nursing--methods--Handbooks.
	Maternity nursing.
	Pediatric nursing.
Form/Genre	Handbooks, manuals, etc.
Notes	Companion for Maternal-child nursing care: optimizing outcomes for mothers, children, and families / Susan L. Ward, Shelton M. Hisley. Philadelphia:F.A. Davis, [2015], ©2016.
	Includes bibliographical references and index.

Therapeutic oligonucleotides: methods and protocols

LCCN	2011931964
Type of material	Book
Main title	Therapeutic oligonucleotides: methods and protocols / edited by John Goodchild.
Published/Created	New York: Humana Press; Springer, c2011.
Description	xi, 340 p.: ill.; 27 cm.
Links	Publisher description http://www.loc.gov/catdir/enhancements/fy1202/2011931964-d.html
	Table of contents only http://www.loc.gov/catdir/enhancements/fy1202/2011931964-t.html
ISBN	1617791873
	9781617791871
	9781617791888
	1617791881
LC classification	RM666.N87 T486 2011
Related names	Goodchild, John (Chemist)
Contents	Therapeutic oligonucleotides / John Goodchild -- Dinucleotides containing 3-S-phosphorothiolate linkages / James W. Gaynor and Richard Cosstick

-- 2-O,4-C-methyleneoxymethylene bridged nucleic acids (2,4-BNACOC) / Yoshiyuki Hari ... [et al.] -- Non-covalent peptide-based strategy for ex vivo and in vivo oligonucleotide delivery / Laurence Crombez ... [et al.] -- Cell-penetrating peptides-based strategies for the delivery of splice redirecting antisense oligonucleotides / Samir El Andaloussi ... [et al.] -- Nanoparticle for tumor targeted delivery of oligomers / Xinrong Liu, Yi Wang, and Donald J. Hnatowich -- Light-directed delivery of nucleic acids / Sigurd Bøe ... [et al.] -- Antibody targeted siRNA delivery / Masoud M. Toloue and Lance P. Ford -- Aptamer-drug conjugation for targeted tumor cell therapy / Michael J. Donovan ... [et al.] -- Five-step process for screening antisense compounds for efficacy: gene target IL-12Rb2 / Nikki B. Marshall, Laura L. Hauck, and Dan V. Mourich -- Diverse small non-coding RNAs in RNA interference pathways / Liande Li and Yi Liu -- Quantification of siRNAs in vitro and in vivo / Angie Cheng, Alexander V. Vlassov, and Susan Magdaleno -- Optimization of transfection conditions and analysis of siRNA potency using real-time PCR / Angie Cheng, Susan Magdaleno, and Alexander V. Vlassov -- siRNA knockdown of gene expression in endothelial cells / Emily Dennstedt and Brad Bryan -- Using RNA interference in Schistosoma mansoni / Rita Bhardwaj, Greice Krautz-Peterson, and Patrick J. Skelly -- Performing the labeled microRNA pull-down (LAMP) assay system: an experimental approach for high-throughput identification of microRNA-target mRNAs / Ren-Jun Hsu and Huai-Jen Tsai -

- Synthesis, purification, and characterization of oligoribonucleotides that act as agonists of TLR7 and/or TLR8 / Tao Lan and Ekambar R. Kandimalla -- Synthesis, purification, and characterization of immune-modulatory oligodeoxynucleotides that act as agonists of toll-like receptor 9 / Mallikarjuna Reddy Putta, Dong Yu, and Ekambar R. Kandimalla -- Surface plasmon resonance investigation of RNA aptamer-RNA ligand interactions / Carmelo Di Primo, Eric Dausse, and Jean-Jacques Toulme -- Practical considerations for analyzing antigene RNAs (agRNAs): RNA immunoprecipitation of argonaute protein / Jacob C. Schwartz and David R. Corey -- Inhibition of human papillomavirus expression using DNAzymes / Maria Luisa Benitez-Hess, Pablo Reyes-Gutierrez, and Luis Marat Alvarez-Salas.

Subjects	Oligonucleotides--Therapeutic use. Oligonucleotides--therapeutic use.
Notes	Includes bibliographical references and index.
Series	Methods in molecular biology, 1064-3745; 764 Springer protocols Methods in molecular biology (Clifton, N.J.); v. 764. Springer protocols.

Tropical infectious diseases: principles, pathogens and practice

LCCN	2014451764
Type of material	Book
Main title	Tropical infectious diseases: principles, pathogens and practice / [edited by] Richard L. Guerrant, MD, Thomas H. Hunter Professor of International Medicine, Director, Center for

	Global Health, Division of Infectious Diseases and International Health, University of Virginia School of Medicine, Charlottesville, VA, USA, David H. Walker, MD, Carmage and Martha Walls distinguished University Chair in Tropical Diseases, Director, Center for Biodefense and Emerging Infectious Diseases, Professor and Chair, Department of Pathology, University of Texas Medical Branch, TX, USA, Peter F. Weller, MD, FACP, FIDSA, Professor of Medicine, Harvard Medical School, Professor, Immunology and Infectious Diseases Department, Harvard School of Public Health, Chief, Infectious Disease Division, Vice Chair of Research, Departhment of Medicine, Beth Israel Deaconess Medical Center, Boston, MA, USA.
Edition	Third edition.
Published/Produced	Edinburgh: Saunders/Elsevier, [2011] ©2011
Description	xxiv, 1130 pages: illustrations (chiefly color); 28 cm.
ISBN	9780702039355 (hbk) 0702039357 (hbk)
LC classification	RC961 .T743 2011
Related names	Guerrant, Richard L. Walker, David H., 1943- Weller, Peter F.
Contents	Principles of Parasitism: Host-Parasite Interactions -- The Perfect Storm: Climate, Ecosystems and Infectious Diseases -- Epidemiology and Biostatistics -- Social and Cultural Factors in Tropical Medicine: Reframing Our Understanding of Disease -- Nutrition and Micronutrients in Tropical Infectious Diseases --

Host Genetics and Susceptibility to Infection -- Disease Eradication and Control -- Vector Biology -- Immunology, Host Defense, Immunodeficiencies, and Vaccines -- Vaccines in the Tropics -- Tropical Infectious Diseases and Malignancy -- Chemotherapy of Parasitic Diseases -- Chemotherapy of Bacterial, Fungal, and Viral Diseases -- Global Surveillance for Emerging Infectious Diseases -- Enteric Escherichia coli Infections -- Typhoid and Paratyphoid (Enteric) Fever -- Nontyphoidal Salmonellosis -- Shigellosis -- Campylobacter Infections -- Cholera Infections -- Noncholera Vibrio Infections -- Enteric Clostridium Infections -- Helicobacter pylori Infections -- Meningococcal Disease -- Gonococcal and Other Neisserial Infections -- Haemophilus ducreyi Infections -- Donovanosis -- Pneumococcal Infections -- Group A Streptococcal and Staphylococcal Infections -- Pertussis -- Legionellosis -- Melioidosis -- Diphtheria -- Tuberculosis and Atypical Mycobacterial Infections -- Mycobacterium ulcerans Infection (Buruli Ulcer) -- Leprosy -- Anthrax -- Bartonelloses -- Brucellosis -- Plague -- Tetanus -- Treponemal Infections -- Relapsing Fever and Other Borrelia Diseases -- Leptospirosis -- Trachoma -- Chlamydia trachomatis Infections of the Genital Tract (Including Lymphogranuloma Venereum) -- Psittacosis -- Spotted Fever Group Rickettsioses -- Typhus Group Rickettsioses -- Scrub Typhus -- Ehrlichioses and Anaplasmosis -- Q Fever -- Measles -- Hendra and Nipah Viral Infections -- Human Herpesvirus Infections --

Smallpox and Related Orthopoxviral Infections -- Respiratory Viral Infections -- Severe Acute Respiratory Syndrome (SARS) -- Enterovirus Infections, Including Poliomyelitis -- Rotavirus Infections -- Calicivirus Infections -- Astroviruses, Enteric Adenoviruses, and Other Gastroenteritis Viral Infections -- Enteric Viral Hepatitis A and E -- Viral Hepatitis C -- Hepatitis B and Deltavirus Infections -- Overview of Viral Hemorrhagic Fevers -- Arenavirus Infections -- Rift Valley Fever -- Crimean-Congo Hemorrhagic Fever -- Hantavirus Infections -- Sandfly Fever, Oropouche Fever, and Other Bunyavirus Infections -- Filovirus Infections -- Yellow Fever -- Dengue and Dengue Hemorrhagic Fever -- Japanese Encephalitis, West Nile, and Other Flavivirus Infections -- Tick-borne Encephalitis and Omsk Hemorrhagic Fever -- Alphavirus Infections -- Rabies -- Human Papillomavirus Infections -- Human Retroviral Infections in the Tropics -- Dermatophytosis -- Mycetoma -- Chromoblastomycosis and Phaeohyphomycosis -- Histoplasmosis, Blastomycosis, Coccidioidomycosis, and Cryptococcosis -- Paracoccidioidomycosis -- Penicilliosis Marneffei -- Mucocutaneous and Deeply Invasive Candidiasis -- Mucormycosis -- Entomophthoramycosis, Lobomycosis, Rhinosporidiosis, and Sporotrichosis -- Pneumocystosis -- Enteric Amebiasis -- Intestinal Flagellate and Ciliate Infections -- Cryptosporidiosis -- Cyclospora, Isospora, and Sarcocystis Infections -- Malaria -- Babesiosis -- African Trypanosomiasis (Sleeping Sickness) --

American Trypanosomiasis (Chagas Disease)* -- Leishmaniasis -- Pathogenic and Opportunistic Free-Living Amebae: Acanthamoeba spp., Balamuthia mandrillaris, Naegleria fowleri, and Sappinia pedata -- Microsporidiosis -- Toxoplasmosis -- Lymphatic Filariasis -- Loiasis and Mansonella Infections -- Onchocerciasis -- Zoonotic Filariasis -- Dracunculiasis -- Toxocariasis and Larva Migrans Syndromes -- Trichinellosis -- Angiostrongyliasis -- Other Tissue Nematode Infections -- Enterobiasis -- Trichuriasis -- Ascariasis -- Hookworm Infections -- Strongyloidiasis -- Introduction to Tapeworm Infections -- Cysticercosis -- Echinococcosis -- Taenia and Other Tapeworm Infections -- Schistosomiasis -- Liver, Lung, and Intestinal Fluke Infections -- Arthropods, Tongue Worms, Leeches, and Arthropod-borne Diseases -- Distinguishing Tropical Infectious Diseases from Bioterrorism -- Health Advice for International Travel -- Migrant, Immigrant, and Refugee Health -- Infectious Diseases in Modern Military Forces* -- Gastrointestinal Symptoms -- Fever and Systemic Symptoms -- Eosinophilia -- Cutaneous Lesions -- Hepatobiliary Disease -- Approach to the Patient in the Tropics with Pulmonary Disease -- Ocular Disease -- Neurologic Disease -- Approach to the Patient in the Tropics with Anemia -- Sexually Transmitted and Urinary Tract Infections -- Approach to the Patient with HIV and Coinfecting Tropical Infectious Diseases -- Delusional Parasitosis -- Tropical Infectious Disease Concerns in Pregnancy.

Subjects Tropical medicine.

Communicable diseases.
Medical microbiology.
Notes "Expert consult"--Cover.
Includes index.

Vaccines & your child: separating fact from fiction
LCCN 2010023588
Type of material Book
Personal name Offit, Paul A.
Main title Vaccines & your child: separating fact from fiction / Paul A. Offit, Charlotte A. Moser.
Published/Created New York: Columbia University Press, c2011.
Description viii, 247 p.; 21 cm.
ISBN 9780231153072 (pbk.)
9780231526715 (electronic)
LC classification RJ240 .O385 2011
Other title Vaccines and your child
Related names Moser, Charlotte A.
Contents Questions parents have about vaccines -- General -- What are vaccines? -- Why do we still need vaccines? -- How do vaccines work? -- How are vaccines made? -- What steps do pharmaceutical companies go through to make vaccines? -- Who recommends vaccines? -- How do we know vaccines work? -- Are vaccine-preventable diseases really that bad? -- Isn't it better to be naturally infected than immunized? -- Are vaccines given in a one-size-fits-all schedule? -- Is there any harm in using an alternative schedule? -- Why can't vaccines be combined to lessen the number of shots? -- Why aren't more vaccines given by mouth? -- Can I avoid vaccines by living a healthy lifestyle? -- Why should I trust a system that makes money for drug companies? -- Should

vaccines be mandated? -- Is it my social responsibility to get vaccines? -- Safety -- Are vaccines safe? -- How do I know if a problem is caused by vaccines? -- What systems are in place to ensure that vaccines are safe? -- How do we know that different vaccines can be given at the same time? -- Do too many vaccines overwhelm the immune system? -- Do too many vaccines weaken the immune system? -- Do too many vaccines cause chronic diseases? -- Do vaccines cause autism? -- Do vaccines cause allergies and asthma? -- Do vaccines cause cancer? -- Do vaccines cause diabetes? -- Do vaccines cause mad-cow disease? -- Do vaccines cause multiple sclerosis? -- Do vaccines cause sudden infant death syndrome (SIDS)? -- Are there "hot lots" of vaccines? -- Is the vaccine adverse events reporting system (VAERS) -- A good way to tell whether a vaccine is harmful? -- Are package inserts useful? -- How do i sort out good from bad information about vaccines? -- Ingredients -- Do vaccines contain products to which children could be allergic? -- Do vaccines contain harmful preservatives like mercury? -- Do vaccines contain harmful adjuvants like aluminum? -- Do vaccines contain harmful chemicals like formaldehyde? -- Do vaccines contain ether or antifreeze? -- Are vaccines made using aborted fetal cells? -- Do vaccines contain products from animals? -- Practical considerations -- How do I deal with my child's fear of shots? -- What can I do to make the vaccine visit less stressful? -- Who shouldn't get vaccines? -- Can I vaccinate my child if he is ill? -- Can I vaccinate my premature

	baby? -- Can I vaccinate my child if he's taking steroids? -- Can I receive a vaccine if I'm pregnant? -- Can I vaccinate my child if I'm breastfeeding? -- What if my child inadvertently got an extra dose of vaccine? -- What if my child inadvertently missed a dose of vaccine? -- Can I vaccinate my child if someone in the home is immune compromised? -- Do vaccines given in countries from which children are adopted still count? -- Individual vaccines -- Vaccines in the first year of life -- Hepatitis B -- Diphtheria,tetanus, and pertussis -- Pneumococcus -- Rotavirus -- Haemophilus influenzae type B -- Polio -- Influenza -- Vaccines in the second year of life -- Measles, mumps, and rubella -- Chickenpox -- Hepatitis A -- Vaccines for adolescents and teens -- Meningococcus -- Human papillomavirus -- The vaccine schedule -- Vaccine schedule and combination vaccines -- Why this schedule? -- Changes to the schedule -- Catch-up schedule.
Notes	Includes bibliographical references and index.

Viral cancers: cytologic tools in diagnosis and management

LCCN	2009039010
Type of material	Book
Personal name	Das, Dilip K. (Dilip Kumar), 1951-
Main title	Viral cancers: cytologic tools in diagnosis and management / Dilip K. Das.
Published/Created	New York: Nova Science, c2010.
Description	xiii, 82 p.: ill. (chiefly col.); 23 cm.
ISBN	9781608764020 (softcover) 1608764028 (softcover)
LC classification	RC268.57 .D37 2010

Contents	Human cancers caused by human papillomavirus and human polyomaviruses: role of cytology -- Role of cytology in human cancers due to Epstein-Barr virus and Kaposi sarcoma herpes virus -- Human cancers following hepatitis B and hepatitis C virus infections: diagnostic role of cytology -- Human cancers caused by human T-cell leukemia virus type 1 and human immunodeficiency virus: role of cytology.
Subjects	Viral carcinogenesis. Cancer--Cytodiagnosis. Tumor Virus Infections--diagnosis. Cytodiagnosis. Early Detection of Cancer.
Notes	Includes bibliographical references and index.
Series	Cancer etiology, diagnosis and treatments series Cancer etiology, diagnosis, and treatments.

Viral oncology: basic science and clinical applications

LCCN	2009036491
Type of material	Book
Main title	Viral oncology: basic science and clinical applications / edited by Kamel Khalili, Kuan-Teh Jeang.
Published/Created	Hoboken, N.J.: Wiley-Blackwell, c2010.
Description	xv, 475 p., [16] p. of plates: ill. (some col.); 25 cm.
ISBN	9780470379912 (cloth) 047037991X (cloth)
LC classification	RC268.57 .V54 2010
Related names	Khalili, Kamel, 1951- Jeang, Kuan-Teh.
Contents	Human papillomavirus-associated cancers / Rachel A. Katzenellenbogin and Denise A.

Galloway -- Molecular events associated with human papillomavirus-induced human cancer / Amy Baldwin and Karl Munger -- The role of the human papillomavirus E6 oncoprotein in malignant progression / Miranda Thomas, David Pim, and Lawrence Banks -- JC virus association with brain tumors: the role of T antigen and insulin like growth factor 1 in DNA repair fidelity / Krzysztof Reiss, Kamel Khalili, and Louis Del Valle -- Involvement of the polyomavirus, JC virus in colorectal cancer / C. Richard Boland, Luigi Ricciardiello, and Ajay Goel -- Possible association of BK virus with prostate cancer / Michael J. Imperiale and Dweepanita Das -- Oncogenic transformation by polyomavirus large T antigen / Abhilasha V. Rathi and James M. Pipas -- Simian virus 40, human infections, and cancer: emerging concepts and causality considerations / Janet S. Butel -- Simian virus 40 and mesothelioma / Natalya Baranova and Michele Carbone -- Molecular immunology of hepatitus B-associated viral cancer / Timothy M. Block and Anand S. Mehta -- Hepititis B vaccine and hepatocellular carcinoma / Mei-Hwei Chang and Ding-Shinn Chen -- Pathogenesis of acute and chronic hepatitus C virus infection / Mark A. Feitelson ... [et al.] -- Molecular mechanisms of hepatitis C virus-induced cellular transformation / Donna Sir and Jing-hsiung James Ou -- Clinical aspects of HTLV-1 associated cancer / Masao Matsuoka -- Oncogenic potential of the HTLV-1 tax protein / Susan J. Marriott -- HIV-associated malignancy: basic and clinical aspects / Melissa Agsalda and Bruce Shiramizu -- HIV-related lymphoma / Giulia De Falco ... [et al.] -- Biology and epidemiology of HHV-8 / Veenu Minhas and Charles Wood -- The role of KSHV in pathogenesis of Kaposi's sarcoma / Gary S. Hayward, Donald J. Alcendor, and Ravit Arav-Boger -- Molecular pathobiology of EBV infection / Joseph Pagano -- Epstein-Barr virus as a pathogen / Wasim A. Dar and Bill Sugden.

Subjects Viral carcinogenesis.
Oncogenic viruses.
Oncogenic Viruses.
Neoplasms--etiology.
Tumor Virus Infections.

Notes Includes bibliographical references and index.

RELATED NOVA PUBLICATIONS

HANDBOOK ON HUMAN PAPILLOMAVIRUS: PREVALENCE, DETECTION AND MANAGEMENT

Harris B. Smith (Editor)

Series: Virology Research Progress, Public Health in the 21st Century
Hardcover ISBN: 978-1-62618-672-9
E-book ISBN: 978-1-62618-673-6
Publication Date: June 2013 566 pages

Human Papillomavirus (HPV) is a common sexually transmitted infection (STI). It affects genital and oral mucosa and is associated with cellular lesions which can progress to cervical cancer and some head and neck cancers respectively. In this handbook, the authors present current research on the prevalence, detection and management of HPV. Topics include HPV in adolescents; clinical utility of HPV testing in cervical cancer screening; HPV epidemiology in females and risk for cervical cancer; HPV infection in pregnant women; immunogenicity of HPV vaccination; HPV in systemic rheumatic disorders; prevalence and distribution of HPV 16, 18 and 58 in Southeast Mexico; the triad of HPV-oral sex-and oral cancer; HPV related malignancies of the reproductive tract; in silico characterization of major capsid protein L1 of HPV type 16 by molecular dynamics and determination of linear B-cell consensus epitopes; the role of HPV in non-small cell lung cancer; and the statistical aspects of HPV modelling in a quantitative manner.

HUMAN PAPILLOMAVIRUS (HPV) AND CERVICAL CANCER HEALTH SYSTEM PREPAREDNESS IN SAINT LUCIA: A POLICY BRIEF*

Krystal Austin, MPH, Anyka Clouden, MPH, Dalia Rassier, MPH, and Praveen Durgampudi†, MBBS, MPH, MSPH

Department of Public Health and Preventative Medicine, St. George's University, Grenada, West Indies

This policy brief focuses on Chronic Disease Preparedness in St. Lucia, focusing specific interventions on cervical cancer and human papillomavirus (HPV) prevention. There is a need to implement a policy to raise awareness of cervical cancer through education and to implement a vaccination program in order to reduce the disease burden of cervical cancer on the St. Lucian population. The policy therefore comprises of two levels. The first is a vaccination campaign targeting girls and boys ages nine to thirteen, in accordance with the World Health Organization guidelines for vaccinating children between the ages of nine and thirteen to protect against HPV infection. The second is through education and screening. There should be implementation of a state wide campaign for the education of the public on

---

* The full version of this chapter can be found in *Public Health Yearbook 2016*, edited by Joav Merrick, published by Nova Science Publishers, Inc, New York, 2017.

† Correspondence: Praveen Durgampudi, MBBS, MPH, MSPH, St. George's University Department of Public Health and Preventive Medicine, St. George, Grenada, West Indies. E-mail: pdurgampudi@sgu.edu.

HPV: what it is, what it causes and how it can be prevented, and routine screening (Pap smears) for women over twenty for the early identification of abnormal cells of the cervix which are a precursor to cervical cancer. With this two level policy, there is hope of reducing the disease burden of cervical cancer on St. Lucia, while raising the awareness of HPV and promoting prevention strategies.

The State-of-the-Art Therapeutic Human Papillomavirus Vaccine*

Ciro Comparetto[1] and Franco Borruto[2]
[1]Division of Obstetrics and Gynecology,
City Hospital, Azienda USL Toscana Centro, Prato, Italy
[2]Consultant in the Field of Health Policy of the
Government of Monaco, Monaco

Cervical cancer is, worldwide, the second cancer in women, with about 500,000 new cases per year and more than 275,000 deaths, of which 88% in developing countries. In Italy, from National cancer registries, there are 3,000 new cases per year, with about 1,500 deaths/year. Screening of women for precancerous lesions is an effective intervention to prevent cervical cancer. This tumor was the first to be recognized by the World Health Organization (WHO) as totally due to an infection: Human Papillomavirus (HPV) is present in 99% of cervical cancers and is considered as their cause. Some types of HPV, the so-called "high-risk" (HR) ones, have been linked to cervical cancer. While most HPV infections reverse their course within a few years, there is no available treatment to accelerate or improve the chances of natural grubbing. The fundamental discovery that HPV infection is the cause of cervical cancer (but also of other cancers, for a total of about 5% of all cancers) opened the way to new primary screening systems (HPV

---

* The full version of this chapter can be found in *Horizons in Cancer Research. Volume 67*, edited by Hiroto S. Watanabe, published by Nova Science Publishers, Inc, New York, 2017.

test) and to new prevention strategies based on vaccination. In fact, although HPV infections are detected with the Papanicolaou (Pap) test, there is not a standard treatment to eradicate infection and to stop the development of precancerous lesions outside the hope that the disease will spontaneously recover. Although HPV vaccination can protect against new cervical HPV infections, women remain at risk for infection by specific haplotypes of HPV that cause cancer and a prophylactic vaccine does generally not help those who have a persistent infection. In addition, both men and women are susceptible to HPV infections in other locations, including the mouth and the anus. Infection can result in a moderate or severe dysplasia (cervical intraepithelial neoplasia – CIN – 2 or 3), which can then be eliminated naturally by the immune system or can progress to cancer.

The current recommended treatment for cervical precancerous lesions is surgical excision, which results in the removal of a portion of the cervix and can cause a weakening of it and the inability to carry a pregnancy to term. A vaccine to prevent or to treat precancerous lesions could represent a way to avoid surgery, which is invasive, and may impair fertility. Recently, two prophylactic vaccines (Gardasil®, produced in yeast, and Cervarix®, produced in insect cells) against the most frequent oncogenic HPV (HPV 16 and 18), capable of conferring efficient protection against persistent HPV infections and consequently from tumors induced by these viruses, have been marketed. The preventive action of both vaccines is due to the administration of "virus-like particles" (VLP), constituted by the assembly of the "late" L1 protein of the viral capsid and highly purified. High production costs (currently about € 180 for the three-dose series) and the need of cold chain for storage, complicate such vaccination in developing countries, where cervical cancer incidence is higher. There is thus a need to develop "second generation" prophylactic vaccines, low-cost, broad-spectrum, cheap, heat-stable, which can be administered in single dose without syringes. At the same time, we need to develop new therapeutic strategies (in particular immune-therapy) for individuals with persistent infections that could not benefit from prophylactic vaccines, obviously. Prophylactic vaccines can prevent some HPV infections, but are not effective against current infections. Unlike prophylactic vaccines that prevent disease, therapeutic vaccines fight the disease after a person is infected in a treatment known as immune-therapy. "Early" oncoproteins E6 and E7 from HR HPV represent ideal candidates for the realization of these therapeutic vaccines, since they are "tumor-specific" antigens. Several experimental vaccines based on these proteins have shown promising results, proving to control or block, even in clinical trials, neoplastic development. As a result, hundreds of millions of already infected women and men (or that will be in the near future) could benefit from this kind of therapeutic intervention.

Preventive and therapeutic vaccines are the two sides of the same coin in the fight against cervical cancer. The first is already available and in many countries, like in Italy, a vaccination campaign in teenagers has started. Preventive vaccine is an extraordinary and very effective weapon against HPV infection and the choice to vaccinate girls before the start of active sexual life will greatly reduce the spread of the virus at least as regards the subtypes against which the vaccine has been developed, which are the most dangerous from the oncological point of view. This represents a solution that combined with early detection through Pap test will allow to achieve important results, but that may not be enough. First, because in many countries of the world there are not the necessary resources available to prevent cervical cancer. Furthermore, preventive vaccination results will be fully visible only after many years, but it is also essential to treat women who are sick today. This is why researchers are working on a therapeutic vaccine. In particular, the new immunological therapies against cervical cancer and other cancers associated with HPV infection at present aim to exploit the features of HPV to activate a cell-mediated immune response against the tumor. It is not easy to activate the immune system against the tumor: the variables are many. The immune system of a woman with cervical cancer is like "asleep" from years of cohabitation with the virus and must be awakened in the right way, wreaking particular immune responses (cell-mediated) that are different from those that are activated with preventive vaccine (antibody-mediated), as evidenced by studies conducted with the support of the Cancer Research Italian Association (AIRC). There is also the need to develop tools to measure the immune response at the mucosal level and in the environment surrounding the tumor, and to discover markers to identify lesions that are most likely to respond to the vaccine. Immunological therapy remains the preferred way for cancers caused by HPV. It is a real scientific challenge for those involved in anti-cancer vaccines. These therapeutic strategies are only just beginning, but we realize their enormous potential.

An experimental vaccine has showed active as a therapeutic treatment for high-grade precancerous cervical lesions caused by HPV in a Phase IIb controlled trial, recently published by *The Lancet*. The vaccine, known for now with the acronym VGX-3100, was the first therapeutic vaccine to be effective against such injuries. Of 107 women who were administered all three doses of the vaccine, 53 (49.5%) achieved a regression of the lesions, documented by biopsy, compared with 11 of 36 (30.6%) in the placebo group. Among women who received at least one dose of the vaccine, 55 of 114 (48.2%) showed a regression of CIN versus 12 of 40 (30%) in the placebo group ($p = 0.034$). VGX-3100, developed by Inovio Pharmaceuticals (Plymouth Meeting, PA, United States of America – US), is a vaccine administered by electroporation, consisting of two synthetic deoxy-ribonucleic (DNA) plasmids that code for two HPV 16 and HPV 18 oncogenic proteins, E6 and E7. The two proteins are presented to immune cells in order to be recognized as part of cancerous and precancerous cells. The vaccine is given by injection in the arm. The most common adverse event was injection-site erythema, slightly more common in the group treated with VGX-3100 compared to placebo (78.4% versus 57.1%). Other common adverse events were fatigue, headache, myalgia, malaise, nausea, and arthralgia. In patients whose biopsies showed complete regression of the lesion, researchers found a significantly higher number of T-cells than in the placebo group ($p = 0.001$). This is a great first step, because it has been shown that the vaccine can trigger an immune response in a person whose immune system initially was not sufficiently activated and had been somewhat impaired, so as to allow the development of the lesion. In their discussion, in addition, study authors say that the results provide promising non-surgical option for the treatment of CIN 2 and CIN 3 and could change the approach to treatment of this very common disease.

Plants represent an alternative platform for the development of anti-HPV both preventive and therapeutic vaccines. It would be, in fact, a good strategy to produce in plants vaccines that represent copies (or biosimilars) of successful (because they are very effective) but expensive products, as the current preventive vaccines against HPV are. Indeed, numerous studies have already shown that the L1 VLP protein can be produced in plants, in a biologically active form, both in a transgenic (tobacco and potato) and in a transient manner. Regarding therapeutic vaccines against HPV-associated diseases, the first demonstration of the possibility to use plants for their production dates back to 2002 when, in cooperation with the Laboratory of Virology of the "Regina Elena" Cancer Institute and the Higher Health Institute (ISS) in Rome, HPV 16 oncoprotein E7 was obtained in tobacco (Nicotiana benthamiana) through a vector derived from the Potato Virus X (PVX). Following immunization of mice with the plant extract, an E7-specific immune response (both antibody- and cell-mediated) was obtained which could inhibit cancer development. These studies have also suggested that the "green vaccine" has intrinsic "adjuvant" (immune-stimulant) properties. Subsequently, the E7 fused protein was expressed at very high yields (3 mg/g fresh tissue) and purified. From infected leaves, an extract used to vaccinate mice was derived. After vaccination in mice, a HPV-associated cancer was induced. Mice resulted protected as for 40% but the tumor grew up in vaccinated mice was much smaller than that present in control mice. This vaccination process has been patented. The system was then enhanced by varying E7 expression strategy in the plant. Results have shown values of effectiveness of 80%. In addition, studies are underway on the use of these plant extracts containing E7 in human dendritic cells for a possible application in autologous vaccinations. These results therefore open the way both to clinical trials on molecules produced in plants and to research and characterization of new active vegetable principles of protein nature for use as immune-stimulants in anti-cancer therapies.

Genticel (Labège-Innopole, France), a biotechnology company and a French detector of innovative immune-therapies to prevent cancers caused by HPV, has developed a new therapeutic vaccine, GTL001, against HPV types 16 and 18, which has been evaluated for its safety in a Phase I clinical trial at the University of Louisville. While prophylactic vaccines are available for those who choose to use them, many women are choosing not to be vaccinated. In addition, most of women in our population are older than vaccine target population, so they cannot have the vaccine and can get HPV infections. Women 25-65 years of age who were diagnosed HPV 16 or HPV 18 have been asked to participate in a small Phase I clinical trial on GTL001 tolerability for a period of only three months. Participants must not have had high-grade squamous intraepithelial lesions (H-SIL) as determined by Pap test. In addition, participants should not have received HPV vaccine and should not have been pregnant or breastfeeding. Participants in the study received two injections at six week of interval as well as various tests and evaluations. All visits, tests, and medicines in connection with the study have been provided at no cost. In addition, participants have been reimbursed for travel expenses. Participants have been registered until the beginning of 2016. Test positions included Louisville, Philadelphia, and Columbus, Ohio.

Genticel today announces positive preclinical in vivo results of the proof of concept of GTL002, its candidate multivalent therapeutic HPV vaccine based on the versatile platform Vaxiclase of the Company. GTL002 tries to eradicate the six most oncogenic types of HPV that affect 158 million women worldwide. GTL001 (known in Europe as ProCervix®), the previous candidate therapeutic vaccine currently in Phase II in Europe, already aimed to prevent infection by the two most oncogenic types of HPV that affect 93 million women. With GTL001 and GTL002, Genticel is the first company that has established a mass pipeline of therapeutic vaccines against HPV for the largest female population charged by this medical unmet need. The proof of concept of GTL002 data showed that an immune response was induced in vivo against each of the six HPV-derived proteins in the therapeutic vaccine. In addition, therapeutic efficacy in vivo has been shown by the extirpation of the tumor in the most widely used and more widely accepted reference model. These positive results, coupled with robust manufacturing data that responded to strict criteria of reproducibility, quality, and performance, provide a solid foundation to advance this multivalent HPV vaccine toward clinical development. Preclinical and clinical pharmacologic, toxicologic, and conventional Good Manufacturing Practice (GMP) Phase III studies have given good results for the preparation and production of Genticel in Europe that could start already in 2017. GTL002 is the first candidate product based on private Vaxiclase technology of Genticel. This next-generation technology platform can be used with many antigens in multiple indications. In February 2015, Genticel signed a first licence agreement with the Serum Institute of India SRL (SIIL), the largest producer in the world of vaccine doses, to evaluate Vaxiclase for use in multivalent vaccines that contain pertussis antigens. Under the agreement, Genticel received up to $ 57 million in advance payments, plus royalties on sales. Reaching the proof of concept of preclinical in vivo effectiveness of GTL002 was an important milestone in strengthening the pipeline of therapeutic HPV Genticel vaccine. These positive data show the ability of researchers to deliver a follow-on candidate product against HPV and support confidence in the substantial value of Vaxiclase in other therapeutic indications.

FACTORS INFLUENCING PARENTAL DECISION MAKING FOR THE HUMAN PAPILLOMAVIRUS (HPV) VACCINE: A LITERATURE REVIEW[*]

Nemica Thavarajah[1,2,•], BSc, Edward Chow[1], MBBS and Jose Arocha[2], PhD

[1]Odette Cancer Centre, Sunnybrook Health Sciences Centre, University of Toronto, Toronto, Ontario, Canada

[2]Faculty of Applied Health Sciences, University of Waterloo, Waterloo, Ontario, Canada

---

[*] The full version of this chapter can be found in *Child and Adolescent Health Yearbook 2015*, edited by Joav Merrick, published by Nova Science Publishers, Inc, New York, 2016.

[•] Correspondence: Nemica Thavarajah, BSc, Department of Radiation Oncology, Odette Cancer Centre, Sunnybrook Health Sciences Centre, 2075 Bayview Avenue, Toronto, ON M4N 3M5, Canada. E-mail: nemica.thavarajah @gmail.com.

The primary objective of this study was to explore the literature to determine common factors that influence parental decision making regarding uptake of the HPV vaccine for their sons and daughters. A systematic literature review was conducted using MEDLINE, EMBASE, and PsychINFO databases to identify studies published between January 2011 and October 2013 reporting factors influencing parental decision making regarding uptake of the HPV vaccine for their children. A total of 19 studies regarding parental decision making for the HPV vaccine for daughters and/or sons and fit the eligibility criteria were retrieved. The sample size of the studies retrieved ranged from 20 to 8652 parents. The most prevalent factors (found in five or more studies) in order of most to least prevalent included knowledge on prophylaxis; concerns with safety and/or side effects; trust in health provider, and/or government, and/or school; sexual health; positive parental attitudes; insufficient information; influence of family and/or friends; and parent-child communication. Many factors influence parental decisions in regards to providing consent for their children to receive the HPV vaccine. Future efforts should focus on addressing these factors to improve uptake of the HPV vaccine by implementing focused strategies and decision aids to help parents make informed decisions.

MECHANISMS OF CANCER FORMATION UNDER INTERPLAY BETWEEN HOST IMMUNITY AND HUMAN PAPILLOMAVIRUS (HPV) INFECTION*

Masachika Senba[1] and Naoki Mori[2]

[1]Department of Pathology, Institute of Tropical Medicine,
Nagasaki University, Nagasaki, Japan
[2]Department of Molecular Virology and Oncology,

* The full version of this chapter can be found in Horizons in Cancer Research. Volume 67, edited by Hiroto S. Watanabe, published by Nova Science Publishers, Inc, New York, 2016.

Gradual School of Medicine, University of the Ryukyus, Okinawa, Japan

The cervical, penil, anal, and oropharyngeal cancers are caused by high-risk human papillomaviruses (HPVs). HPVs have developed strategies to escape host defense by innate and adaptive immunity. HPVs can induce immune evasion of the infected cells, which enable the virus to be undetectable for long time. Immune response evasion has been considered significant aspect of HPVs persistence, which is the principal contributing factor leading to HPV associated with cancers. HPVs have evolved in a different way of evading immune system. The macrophages and natural killer (NK) cells are classified as innate immune cells, and dendritic cells (DCs), cluster of differentiation $CD4^{+}/CD8^{+}$ T-cells and regulatory T-cells are classified as adaptive immune cells. HPV associated with cancers expressing viral oncogenes E6 and E7 are potentially recognized by the immune system. The major histocompatibility complex (MHC) class I molecules are patrolled by NK cells and $CD8^{+}$ cytotoxic T lymphocytes. This recognition system is a main target for the strategies of immune evasion deployed by viruses. The long latency period from initial infection to persistence signifies that HPV evolves mechanisms to escape the immune response. The integration of the viral genome into the host cells results in induced cellular immortalization, transformation, and carcinogenesis, due to their interference with cell cycle and apoptosis control. Most importantly, high-risk HPV E6 and E7 oncoproteins are able to bind and degrade tumor suppressor p53 and pRb genes, respectively. Therefore, interaction of p53 and pRb proteins can give rise to an increased genomic instability following induced immortalization, transformation, and carcinogenesis. Toll-like receptors (TLRs) are specific pattern recognition molecules that bind to HPV components to trigger innate and adaptive immunity. HPVs modulate TLRs expression and interfere with TLRs signaling pathways, lead to persist viral infection and oncogenesis. Overexpression of NF-κB in HPVs associated with cancers suggests that NF-κB activation is a key modulator in driving chronic inflammation to cancer. HPVs oncogene-mediated suppression of NF-κB activity contributes to HPV escape from the immune system. This review focuses on the molecular players of the HPVs infection lead to chronic inflammation and cancer.

Utilization of Peer Teen Advocates to Increase Awareness of Human Papillomavirus and Vaccination among Urban Youth*

Denise Uyar[1, †], MD, Staci Young[2], PhD, Sarah Tyree-Francis[3] CFLE and Mary Ann Kiepczynski[4], RN

[1]Medical College of Wisconsin, Department of Obstetrics and Gynecology,
Division of Gynecology Oncology, Wisconsin US
[2]Medical College of Wisconsin,
Department of Family and Community Medicine, Wisconsin US
[3]Program Manager, Adolescent Holistic Health Initiatives,
Boys & Girls Clubs of Greater Milwaukee, Wisconsin US
[4]City of Milwaukee Health Department, Disease Control & Environmental Health,
Public Health Nurse Immunization Program Coordinator, Milwaukee, Wisconsin US

The city of Milwaukee, Wisconsin ranks among the highest urban areas for sexually transmitted disease in the United States with human papillomavirus being one of the most common sexually transmitted infections. In 2006, the Advisory Committee on Immunization Practices

---

* The full version of this chapter can be found in *Public Health Yearbook 2015*, edited by Joav Merrick, published by Nova Science Publishers, Inc, New York, 2016.

† Corresponding author: : Denise Uyar, MD Medical College of Wisconsin, Associate Professor Department of Obstetrics and Gynecology, Division of Gynecology Oncology 9200 West Wisconsin Avenue, Milwaukee, WI 53226, United States. E-mail: duyar@mcw.edu.

recommended routine vaccination of adolescent girls at ages 11 or 12 years with 3 doses of the Food and Drug Administration approved HPV vaccination. Despite the availability, safety and demonstrated efficacy of these vaccines, only 33% of adolescent girls in the United States have received all three doses of the series. In Milwaukee, less than 30% of adolescents have received the vaccination series. Several studies have evaluated the barriers for HPV vaccination. Consistently, deficits in knowledge and education for both parents and adolescents regarding HPV and the vaccination are cited as reasons why youth did not get the vaccine. This study was created to determine if today's urban youth could successfully be engaged in a program of positive health messaging surrounding HPV and the HPV vaccination using peer educators and social media.

Human Papillomavirus (HPV) and Cervical Cancer Prevention*

Krystal Austin, MPH, Anyka Clouden, MPH, Dalia Rassier, MPH and Praveen Durgampudi†, MBBS, MPH, MSPH

Department of Public Health and Preventive Medicine,
St. George's University, Grenada, West Indies

In this chapter we focus on Chronic Disease Preparedness in St. Lucia, focusing specific interventions on cervical cancer and human papillomavirus (HPV) prevention. There is an urgent need to implement a policy to raise

* The full version of this chapter can be found in *Caribbean Adolescents: Some Public Health Concerns*, edited by Cecilia Hegamin-Younger and Joav Merrick, published by Nova Science Publishers, Inc, New York, 2015.

† Correspondence: Praveen Durgampudi, MBBS, MPH, MSPH, St. George's University Department of Public Health and Preventive Medicine, St. George, Grenada, West Indies. E-mail: pdurgampudi@sgu.edu

awareness of cervical cancer through education and to implement a vaccination program in order to reduce the disease burden of cervical cancer on the St. Lucian population. The policy therefore comprises of two levels. The first is a vaccination campaign targeting girls and boys ages nine to thirteen, in accordance with the World Health Organization guidelines for vaccinating children between the ages of nine and thirteen to protect against HPV infection. The second is through education and screening. There should be implementation of a state wide campaign for the education of the public on HPV: what it is, what it causes and how it can be prevented, and routine Pap smear screening for women over twenty for the early identification of abnormal cells of the cervix which are a precursor to cervical cancer. This two level policy, creates an opportunity of reducing the disease burden of cervical cancer on St. Lucia, while raising the awareness of HPV and promoting prevention strategies.

Human Papillomavirus in Head and Neck Cancer: Detection Methods and Clinical Significance*

Keziah John[1], Maryam Jessri[1] and Camile S Farah[1,2†]
[1]Oral Oncology Research Program, UQ Centre for Clinical Research, The University of Queensland, Herston, Queensland, Australia
[2]The Australian Centre for Oral Oncology Research & Education, Brisbane, Queensland, Australia

Head and neck squamous cell carcinoma (HNSCC), the sixth most common cancer in the world, is associated with risk factors such as the use

* The full version of this chapter can be found in *Head and Neck Cancer: Epidemiology, Management and Treatment Outcomes*, edited by Madison R. Hall, published by Nova Science Publishers, Inc, New York, 2015.

† Corresponding author: Professor Camile S Farah, The Australian Centre for Oral Oncology Research & Education (ACORE), PO Box 88, Royal Brisbane & Women's Hospital, Herston QLD 4029 Australia, Tel: +61 7 3012 8760, Fax: +61 7 3012 7960, Email: camile@oralmedpath.com.au

of tobacco, alcohol, areca nut, and more recently apparent, chronic infection of human papillomavirus (HPV). HPV-positive HNSCC is associated with improved clinical outcomes despite presentation at more advanced stages. This article reviews the contribution of HPV to HNSCC development and tumorigenesis, and the current literature on HPV detection methods. It also discusses the role of microRNAs and epigenetic changes involved in HPV-positive tumours, its response to radiation therapy, and a discussion on prognosis.

PEER TEEN ADVOCATES AND INCREASED AWARENESS OF HUMAN PAPILLOMAVIRUS AND VACCINATION AMONG URBAN YOUTH*

Denise Uyar[1†], Staci Young[2], Sarah Tyree-Francis[3] and Mary Ann Kiepczynski[4]

[1]Medical College of Wisconsin Department of Obstetrics and Gynecology, Division of Gynecology Oncology,
[2]Medical College of Wisconsin, Department of Family and Community Medicine,
[3]Adolescent Holistic Health Initiatives, Don and Sallie Davis Boys and Girls Club and
[4]City of Milwaukee Health Department, Disease Control & Environmental Health, Public Health Nurse Immunization Program Coordinator, Milwaukee, Wisconsin, United States of America

---

* The full version of this chapter can be found in *Public Health: Improving Health via Inter-Professional Collaborations*, edited by Rosemary M. Caron and Joav Merrick, published by Nova Science Publishers, Inc, New York, 2014.

† Corresponding author: Denise Uyar, MD Medical College of Wisconsin, Associate Professor Department of Obstetrics and Gynecology, Division of Gynecology Oncology 9200 West Wisconsin Avenue, Milwaukee, WI 53226, United States. E-mail: duyar@mcw.edu.

The city of Milwaukee, Wisconsin ranks among the highest urban areas for sexually transmitted disease in the United States with human papillomavirus being one of the most common sexually transmitted infections. In 2006, the Advisory Committee on Immunization Practices recommended routine vaccination of adolescent girls at ages 11 or 12 years with 3 doses of the Food and Drug Administration approved HPV vaccination. Despite the availability, safety and demonstrated efficacy of these vaccines, only 33% of adolescent girls in the United States have received all three doses of the series. In Milwaukee, less than 30% of adolescents have received the vaccination series. Several studies have evaluated the barriers for HPV vaccination. Consistently, deficits in knowledge and education for both parents and adolescents regarding HPV and the vaccination are cited as reasons why youth did not get the vaccine. This study was created to determine if today's urban youth could successfully be engaged in a program of positive health messaging surrounding HPV and the HPV vaccination using peer educators and social media.

GENETIC DIVERSITY OF HUMAN AND ANIMAL PAPILLOMAVIRUSES*

Antonio Carlos de Freitas,[a†] Ana Pavla Almeida Diniz Gurgel,[a] Bárbara Simas Chagas,[a] Valdir de Queiroz Balbino[b] and Marcus Vinicius Aragão Batista[b]

[a]Laboratory of Molecular Studies and Experimental Therapy,

* The full version of this chapter can be found in *Advances in Genetics Research. Volume 12*, edited by Kevin V. Urbano, published by Nova Science Publishers, Inc, New York, 2014.

† Corresponding author: Antonio Carlos de Freitas, Department of Genetics, Federal University of Pernambuco, Cidade Universitária, CEP 50.670-901, Recife – Pernambuco – Brazil. E-mail: acf_ufpe@yahoo.com.br.

Department of Genetics, Center for Biological Sciences,
Federal University of Pernambuco, Pernambuco, Brazil
[b]Laboratory of Bioinformatics and Evolutionary Biology,
Department of Genetics, Center for Biological Sciences,
Federal University of Pernambuco, Pernambuco, Brazil

The papillomaviruses form a highly diverse group that infects mammals, birds and reptiles. There is a huge veterinary and public health importance associated with these viruses. Because of this, human and bovine papillomaviruses are the most studied viruses of this group. Infection with human papillomavirus (HPV) represents the most important risk factor in the genesis of cervical cancer. Bovine papillomavirus (BPV) infects basal epithelial cells leading to the formation of benign and malignant tumors, when in synergism with action of environmental co-factors. Brazil is a country with continental dimensions; therefore the understanding of the genetic variability and the distribution of these viruses is essential for an active surveillance program and the design of more efficient vaccines and diagnostic methods. In what concerns the association of HPV and cervical cancer, HPV-16 is the most frequent type worldwide, with the exception of Indonesia where the HPV-18 happens to be the most common type. Although the HPV-16 is the most frequent, the second most important type differs between different regions in the world. So, this difference in the incidence of HPV genotypes needs to be elucidated, along with the genetic variability of these isolates that could lead to distinct oncogenic potential among different variants and subtypes. Oncogenic variants of HPV-31 were found and some of the mutations were observed in regions involved in host immune recognition. Changes in amino acid sequences between variants from different geographical regions evidence the possibility of functional divergence between them, suggesting that the variants detected may have different adaptive fitness. On the other hand, twelve BPV types are known so far, but only three of them are associated with cancer. There is a great diversity of BPV types among cattle and high co-infection was found. The presence of infection with multiple viral types could lead to suppression of the immunological system making the clinical state of the animals worse.

So, further molecular epidemiological investigations on the incidence and diversity of papillomavirus infection will assist in establishing a more accurate view of the distribution of these viruses. This could help the development of accurate and specific diagnostics, prophylactics and therapeutic requirements.

Human Papillomavirus: Factors Influencing Vaccination Decisions among Hispanic Parents with Adolescent Daughters*

Mindy Ma[1] and Jeffrey L. Kibler[2]
[1] Division of Social and Behavioral Sciences,
Farquhar College of Arts and Sciences
Nova Southeastern University, Fort Lauderdale, FL, US
[2] Center for Psychological Studies
Nova Southeastern University, Fort Lauderdale, FL, US

Human Papillomavirus (HPV) is known to cause 70% of cervical cancer. Latina women in the United States are disproportionally impacted by cervical cancer. The US Food and Drug Administration recently licensed two vaccines to protect against HPV. The objective of this chapter is to review the existing data on HPV vaccination and factors contributing to Hispanic parents' decision to vaccinate their daughters.

* The full version of this chapter can be found in *The Psychology of Women: Diverse Perspectives from the Modern World*, edited by Jamie Marich, published by Nova Science Publishers, Inc, New York, 2012.

Hyperthermia Therapy in Muco-Cutaneous Infection by Human Papillomavirus: Efficacy and Mechanisms of Action*

Xing-Hua Gao, Wei Huo, Yuxiao Hong and Hong-Duo Chen
Department of Dermatology,
No. 1 Hospital of China Medical University

Human papillomavirus (HPV) infection of the skin or mucosa leads to various clinical forms of warts, or even carcinomas. Most of the warts will regress in a matter of years, while some are recalcitrant. Warts are disfiguring or debilitating. There is quite an array of physical, chemical, pharmaceutical or biological therapies for treating warts. Most of the currently applied methods have certain limitations, such as low cure rates, ablation related sufferings, long healing time. Local heat therapy, or local hyperthermia within or several degrees above fever range, has been anecdotally practiced in treating warts, while high degree evidence based studies are not abundant. In the present review, we collected and analyzed literatures on hyperthermia treatment of warts, in the aim to best evaluate its efficacy. The natural history of wart suggests a systemic immune response to resolve the HPV infection. HPV generally infects epidermal compartment, It is thus reasonable to postulate that two major epidermal components, keratinocytes-- the body of HPV infection, and Langerhans cells-- professional antigen presenting cells, take principal roles in initiating and regulating the immune responses against HPV. Here we reviewed the effect of hyperthermia on immune related responses in keratinocytes and

* The full version of this chapter can be found in *Hyperthermia: Recognition, Prevention and Treatment*, edited by Xing-Hua Gao, published by Nova Science Publishers, Inc, New York, 2011.

Langerhans cells, at both cellular and molecular levels. Last but not the least, the effect of hyperthermia on the biology of HPV itself is also discussed.

INDEX

A

B

C

D

E

F

P

R

S

T

U

V

W